PETERSEN'S BOOK OF MAN IN SPACE

BEYOND THE THRESHOLD

Earthrise seen from lunar orbit

Volume Five

MAN IN SPACE Vol. 5 BEYOND THE THRESHOLD

Edited by Al Hall and the General Subjects editors of Specialty Publications Division.

COVER

The Lunar Module *Eagle* hovers over man's newest conquest after Apollo 11's historic first landing on the moon. The earth shines brightly in the background. Photo by Apollo astronaut Michael Collins. Cover design by Pat Taketa.

INSIDE FRONT COVER

Nations around the world commemorate the efforts of the United States and the Soviet Union in space exploration with postage stamps. (From the collection of Kurt Raymond Hall.)
Library of Congress Catalog Card Number 74-81003 ISBN 0-8227-0077-8

The editors of Petersen's MAN IN SPACE are indebted to the following agencies, companies, and individuals for their enthusiastic cooperation in helping track down and make available the numerous photos and hundreds of bits of information which add immeasurably to the scope and content of this volume. Our profuse thanks to: National Aeronautics and Space Administration (who supplied all photographs unless otherwise credited), and especially Les Gaver, Margaret Ware, and the Audio-Visual Branch, NASA Headquarters; Jack Stewart, Karla Scott, and NASA's Ames Research Center; Brian Erb, John MacLeish, and NASA's Johnson Space Center; Barbara Rogers and NASA's Kennedy Space Center; Joe Jones and NASA's Marshall Space Flight Center; Warren Tosh and the Aerospace Systems Division of the Bendix Corp.; James D. Grafton and the Boeing Aerospace Company; Mr. E. Russell, Jr. and the IBM Corp.; Park H. Irvine and the Northrop Corp.; Walter Braun, Elsie Behmer, and Paillard, Inc.; Joyce Lincoln and Rockwell International's Rocketdyne Division; Anthony J. Longo and Rockwell International's Space Division; Mike Gentry and the Technicolor Corp.; Don Bane and TRW Systems Group, Inc. And our thanks to the 65 astronauts, the thousands of companies, and the hundreds of thousands of people who worked together to make possible man's greatest journey.

CONTENTS

INTRODUCTION

In the billion-year span of history, it was but a brief moment in time. Yet the world's population was more aware of that 194 hours, 24 minutes of Apollo 11 than any comparable period in the history of man. Never before had so much of mankind been offered the opportunity to be an eye-witness to a momentous event *as it was happening!* This alone makes comparison between Apollo 11's landing on another celestial body and any other historic first virtually impossible. When Columbus discovered the new world, news of his adventures was made known only after his return, and then it was to a small audience. Likewise, when Admiral Peary dashed to the North Pole in 1909, news of his epic journey was delayed, if only for a short time.

The same near-miraculous technology that enabled America to achieve a moon landing also provided the unheard-of opportunity of making every American who cared to look and listen a front-row spectator to this historical happening. Thanks to the vast and incredibly complex NASA communications network and the numerous satellites which preceded man into space, millions upon millions of Americans were joined in their televiewing by other eye-witnesses in countries around the world in seeing man's first lunar step.

But how many of those who were privy to peer over Armstrong's and Aldrin's shoulders, or quietly contemplate the overwhelming feeling of loneliness experienced by Michael Collins in "spaceship Columbia", actually took advantage of the unparalleled television and radio coverage? And of those who did, how many actually realized what was happening or understood the significance of what they heard?

And what of the missions that followed? They are the subject of this volume. The post-Apollo 11 missions enjoyed superior picture quality as compared to earlier telecast missions, but suffered by way of quantity. Was it, as network executives said, because the American public was losing enthusiasm over the space program? Or was the scant coverage accorded to these later missions simply due to an overabundance of material the television and radio media thought was "more interesting?" Material such as wars, murders, politics and epic disasters were in great abundance during the period from late 1969 through early 1973, the era covered by this volume of MAN IN SPACE.

BATTLE FOR SPACE

No, scientific explorations and significant discoveries (even if they were in space and on the moon) were hard pressed to compete with violence and intrigue for equal time and space in the media. Apollo 11 did not cause the world to stop (perhaps it paused a bit, but it certainly didn't stop), and later missions could surely not be expected to accomplish this task either. Certainly not in the face of the headline news pouring forth almost daily from Viet Nam; Ulster, Ireland; East Pakistan; Washington, D.C.; and the Suez Canal. Man's journey beyond the first small step of exploration of the moon to its detailed investigation was forced to share the world's attention with myriad other happenings. It might be worthwhile to place the events that are detailed in this volume against the backdrop of world events for the sake of a better perspective.

WARS

Most of the attention was still focused on the Vietnam conflict. The United States had been actively involved in Indo-China for nearly eight years, although hope was held out for a peace settlement when President Nixon announced that 30,000 of the nearly 534,200 American troops would be withdrawn by August 31, 1969. Additional troop withdrawals according to a secret timetable drawn up by the President continued beyond the time span covered by this volume, but did little during the aforementioned period to bring peace between South and North Vietnam. Even the death of Ho Chi Minh, President of North Vietnam, on September 3, 1969 did not move either side closer to settlement. All the while, U.S. negotiators continue to hold sporadic peace talks with the North Vietnamese in Paris. Early 1970 saw the Vietnam war spill over into neighboring Cambodia and stepped-up bombing raids on North Vietnam targets by U.S. bombers. Then, in April 1970, President Nixon committed U.S. troops to attack communist strongholds along the Cambodian border and up to 20 miles inland in some areas. This was done in cooperation with South Vietnamese troops and at the request of Cambodia's government. The American troops were withdrawn within two months, and the South Vietnamese continued with the offensive. In September 1970, American B-52 bombers pounded a sparsely settled area along the North Vietnam-Laotian border where communist troops were suspected of gathering for a mass offensive. The following year, 1971, saw South Vietnamese engage the communists near the Laotian border, and in '72, the U.S. said it had 50,000 troops stationed in Thailand. Thus, while the number of American troops grew smaller, the battleground expanded.

Don't think for a minute, though, that the Vietnam conflict was the only war vying for the news media's attention. No indeed. In August of 1969, British troops entered Ulster, Ireland, to try and bring an end to the fighting between Northern Ireland's Protestants and Catholics. That battle also continues beyond the era spanned by this volume in a hail of sniper fire, bombings and murders. The same holds true for the fighting along the Suez Canal between Arabs and Israelis.

In 1970, a national election (the first since gaining independence from Britain) split the Islamic nation of Pakistan into two factions: West Pakistan (Punjabi) and East Pakistan (Bengali). In the war that followed, India stepped in after provocation and the result became the new Bengali nation of Bangladesh. It has been estimated that the two-year struggle between East and West Pakistan cost upwards of 200,000 lives.

But it was not nearly as costly a war as Biafra's struggle to break from the domination of Nigeria. This ended in failure in January 1970, after 31 months of civil war. Only a massive international relief operation saved nearly one million Ibo tribesmen from certain starvation.

The early '70s also saw airplane hijackings grow to almost epidemic proportions. However, there was an added twist. Instead of the five words pilots had learned to dread ("Fly this plane to Cuba"), there came requests for parachutes and very large sums of cash money. The first act of air "piracy" was credited to someone known only as D.B. Cooper, who skyjacked a Boeing 727 jetliner, demanded and got $200,000 in ransom for the plane and crew, then parachuted from the rear door into the wilds of the Pacific Northwest. This new form of extortion against airlines proliferated to the point that strict security measures became commonplace at airports, even though the skyjackers were usually nabbed and the ransom money recovered without injuries or death. Nevertheless, each and every skyjacking received instant and continued coverage by the news media.

MURDERS

Murder, as always, rates the front page and prime time. This era had its share and perhaps a little more. Campus violence reached a peak in May of 1970, when four students were killed by National Guardsmen at Kent State University in Ohio, and two Jackson State students were killed in an exchange of gunfire between a sniper and Mississippi police. On March 29, 1971, 1st Lt William Calley was found guilty of the premeditated murder of at least 22 South Vietnamese civilians in the village of My Lai. In California, Charles Manson and his "hippie" clan were convicted of the cold-blooded murders of Sharon Tate, four of her friends, and Los Angeles businessman Leno LaBianca and his wife. Also in California, farm labor contractor Juan Corona was indicted for murder after police discovered the bodies of 23 transients buried in orchards by Yuba City. But the gruesome peak occurred during the 1972 Olympic Games in Munich, Germany, when Arab terrorists entered the Israeli athletes' quarters and, after shooting two who attempted to block their entry, tried to hold the remaining nine as hostages to gain the release of terrorists held in Israeli jails and their own safe passage out of Germany. Most of the world watched the drama on television, relayed via satellite, as the terrorist attempt ended with the deaths of 11 Israeli Olympic Team members and five of the eight Arab terrorists.

DISASTERS

Other headlines during the era under discussion told of floods in Guatemala, South Korea, Algeria, Tunisia, Brazil, Romania and India. There were also earthquakes in Greece, Yugoslavia, Manila, Los Angeles, Turkey, and Peru. The latter left 50,000 dead and 200,000 injured.

POLITICS

Politics, a subject always good for a few laughs, was not very humorous in the early '70s. United States Presidential candidate and Governor of Alabama George Wallace was paralyzed and forced to withdraw from the campaign when he became the victim of an assassination attempt after a speech in a Silver Springs, Maryland, shopping center. Democratic candidate George McGovern had problems of another kind. Assured of the Democratic nomination, he found it difficult to find a running mate after disclosure of the fact that Thomas Eagleton, his first vice-presidential choice, had been hospitalized for nervous exhaustion and had undergone electroshock therapy. Seventeen days after he was nominated, Eagleton was dropped from the Democratic ticket, to be later replaced by R. Sargent Shriver. McGovern and Shriver lost the '72 presidential election to Richard Nixon and Spiro Agnew. Naturally, as in any election, there were claims and counterclaims, charges and countercharges. But the one that eventually proved most damaging was the break-in and bugging of the Democratic national headquarters in the Watergate Hotel, thought to be masterminded by members of Nixon's campaign committee.

BRIGHTER SIDE

All the news of the era under discussion was not bad, but there were very few bright spots. One of them came as the result of an invitation extended by Red China to host a U.S. table tennis team on a tour of mainland China. The team was personally greeted by Communist Chinese Premier Chou En-Lai in April of 1971, and paved the way for a visit to Red China by President Nixon in February 1972. The press credited "ping pong diplomacy" for easing the tensions between the U.S. and mainland China. Three months later, President Nixon paid a personal visit to the Soviet Union, and during talks with Communist leaders signed an agreement on a U.S.-U.S.S.R. venture in space exploration aimed at developing compatible docking sys-

The Chinese won at ping pong, but the Russians lost at chess, as American Bobby Fischer defeated Boris Spassky and won the world chess championship. Recluse billionaire Howard Hughes defeated attempts by Clifford Irving and wife, Edith, to swindle McGraw-Hill Publishers out of $750,000 for an "autobiography" of Hughes, when he held an open phone conversation denouncing the Irvings and their "book" and denied ever meeting Irving, much less talking with him.

OUR APPROACH

Yes, all that's been mentioned and much, much more filled the newspapers, television screens, and radios in the early Seventies. The post-Apollo 11 missions had a tough act to follow, but they bravely went on stage for their moment of glory. To those of you who saw but did not quite comprehend, and who listened but didn't quite understand, and to you who were not old enough to do either, we dedicate this volume of MAN IN SPACE. On the following pages is the complete chronology of man's steps beyond the threshold of lunar exploration. You'll find the material written as though it were happening now. The basic reason for this is that we are not yet far enough removed from the actual events to place them in true historical perspective. That task will be left to generations not yet born. And at that time, questions may arise that perhaps this volume will answer. For this reason alone "Beyond The Threshold" (like the other volumes in this series) is written in the present tense even though the incidents are already history. There is another, and perhaps more important, reason for a narrative in this style: it allows those who never experienced the excitement and drama to catch a glimpse of what it was like to be present at the birth of a new era. It also permits those who looked and did not see or listened and did not hear a chance to relive what may well prove to be the most exciting age in mankind's history.

CHANGES

Youngsters of today see a different moon than that seen by their elders. Thanks to Project Apollo, it is almost as familiar to them as their own neighborhood, and those who were not yet born or who were too young to understand what all the fuss was about on that 20th day of July in Nineteen Hundred Sixty-nine will never experience the thrill of lying atop a grassy knoll on a warm summer night and gazing in awe at a cold, glowing moon and wondering. Wondering what it would be like to go there? Wondering what mysterious things one might find if they could but look at the other side? Wondering if perhaps they might encounter strange creatures lurking and waiting? Wondering if there is really a way to reach that bright and distant orb?

Fortunately, there are other places out there in the cold darkness of space that beg for our attention, and those who missed the thrill of one first in space may be eye-witnesses to another. This book is not the end. Other things are happening and man's curiosity will drive him to greater discoveries. On page 142, Dr. Fletcher, Administrator of the National Aeronautics and Space Administration, previews what is in store for the future of our space program. It offers a great deal of promise for the younger generation.

THE EDITORS

APOLLO 12

For the second manned landing NASA selects a site in the Ocean of Storms (Oceanus Procellarum). The spot desired is about 200 miles south of the great crater Copernicus, and about 50 miles north of the Known Sea (Mare Cognitum). It is not an unexplored region; Ranger 7 impacted in the Mare Cognitum on July 31, 1964; the Soviet Luna 5 crashed about 30 miles northwest of the site on May 12, 1965; and Surveyor 3 successfully soft-landed on April 20, 1967. Apollo 12 will attempt to find and land right beside the Surveyor. Apollo 12 will also deploy the first ALSEP (Apollo Lunar Surface Experimental Package), an integrated array of experiments powered by a tiny nuclear reactor. Planned are two EVAs, each lasting over 3½ hours. Since the landing site is over a thousand miles from the Apollo 11 site, it is hoped a different type of rock will be encountered. The target launch date is November 14, 1969. For this mission of precise navigation, NASA calls upon a trio of Navy captains, who naturally select nautical names for the modules: the CM becomes *Yankee Clipper,* and the LM is christened *Intrepid.* The mission emblem has a delightful fairy-tale appearance, depicting an old-time clipper ship sailing the seas of space to the moon.

CHARLES CONRAD, JR.

SPACECRAFT COMMANDER/This is the third space flight for Navy Captain Conrad, one of the most experienced astronauts. The former Philadelphian and Princeton graduate was the pilot on the Gemini 5 flight, and command pilot on Gemini 11, when he was first teamed with Gordon. Subsequently the two were paired again in the backup crew for Apollo 9, Conrad as Spacecraft Commander. A devotee of aquatic sports and auto racing, Conrad is the father of four children. His backup this time is USAF Col. David R. Scott.

RICHARD F. GORDON, JR.

COMMAND MODULE PILOT/A Navy Captain born and reared close to the sea in Seattle, Gordon graduated from the University of Washington in 1951. He was one of the Navy's hottest test pilots, logging 3500 jet hours in fighters such as the F8U, F11F, and A4D, and was the first project test pilot for the F4H Phantom II. He won the 1961 Bendix Trophy Race. This is his second space flight, having been the pilot on Gemini 11; he was backup CM pilot on Apollo 9. Backing him on this flight is Lt. Col. Al Worden, USAF.

ALAN L. BEAN

LUNAR MODULE PILOT/Born 37 years ago in Wheeler, Texas, Captain Bean received a BS degree in Aeronautical Engineering from the University of Texas. Like his crewmates, he was a test pilot at the Naval Air Test Center at Patuxent River, Maryland, and logged over 4700 hours in 27 different types of military aircraft. Although a space rookie, he was backup command pilot for Gemini 10 and backup LM pilot for Apollo 9. He is dad to two children, and relaxes with swimming and painting. His backup is Col. Jim Irwin, USAF.

APOLLO XII
Conrad • Gordon • Bean

APOLLO 12 Preparation

As the launch date draws inexorably closer, the months of general training for Apollo 12's crew turn into weeks of intensive specialized preparations. Each member will have specific jobs to perform, and must know also those of his fellow crew members in case an emergency should arise. Aboard the Command Module Simulator, (**1**), CM pilot Richard Gordon learns the intricacies of the Hasselblad lunar camera equipped with a Zeiss 500mm telephoto lens. On the mission commander's couch in the foreground is more photographic equipment. At Ellington Air Force Base, Texas, mission commander Charles Conrad takes a spin in the Lunar Landing Test Vehicle (**2**). Dubbed the "flying bedstead", the LLTV was used to train both the mission commander and Lunar Module pilot in lunar landing techniques. But astronauts are not the only ones who practice their complex tasks. At Cape Kennedy, the team that will launch Apollo 12 rehearses with a "countdown demonstration test", a

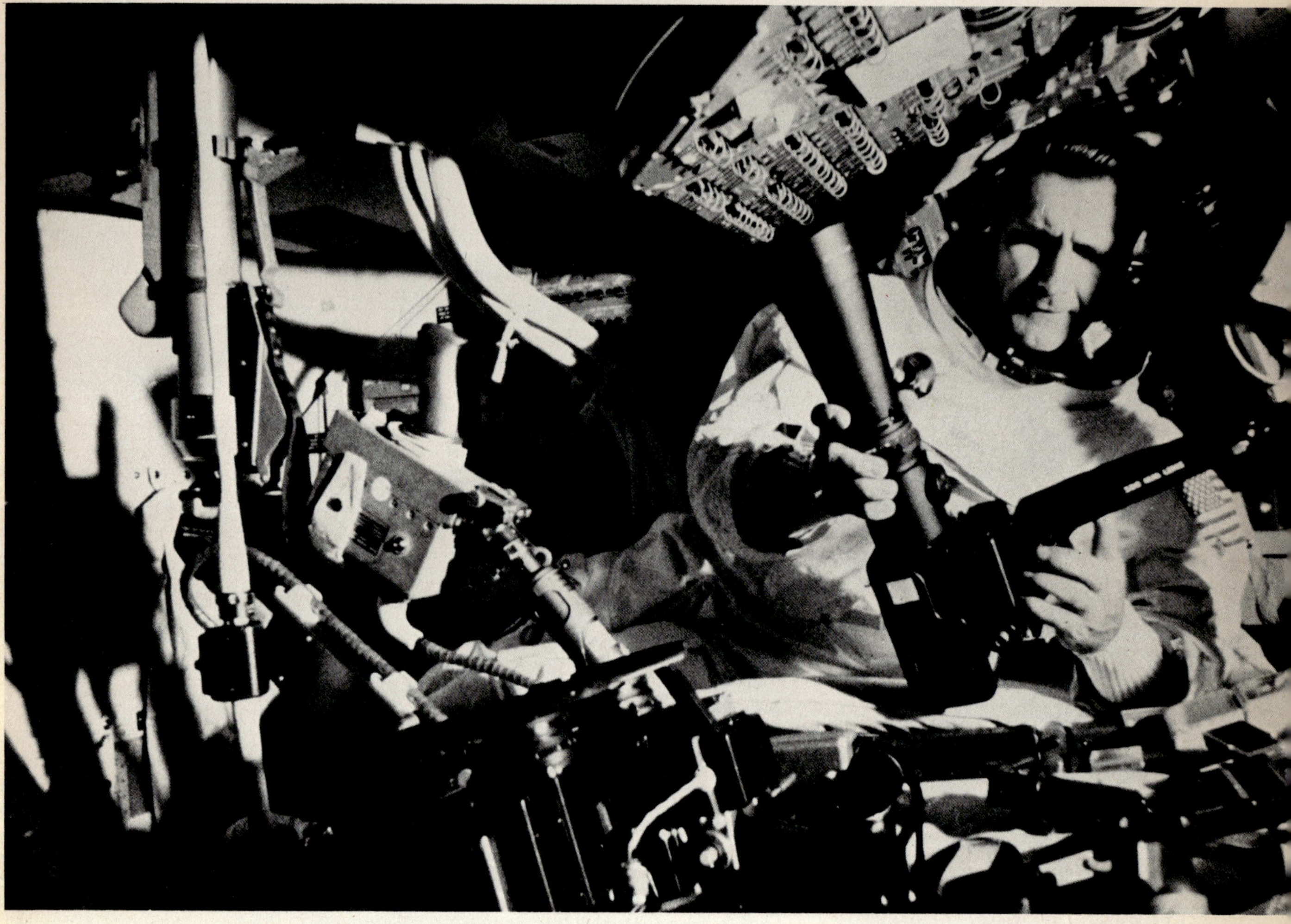
1

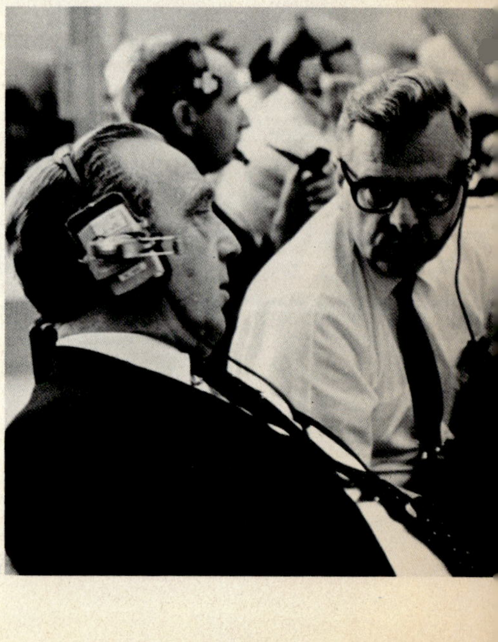
3

dry run of the actual launch (**3**). From Firing Room 2 of the Launch Control Center, Dr. Kurt Debus (foreground), Director of the Kennedy Space Center, discusses an aspect of Apollo 12's CDDT with Walter Kapryan, Director of Launch Operations. (**4**) A fisheye lens captures the crew as they rehearse their flight in the CM simulator. As spacecraft pilot, Gordon sits in the center, flanked by LM pilot Alan Bean (left) and mission commander Charles Conrad. Gordon's right hand is on the main rotation and translation controller, while Conrad is touching another flight controller. To acquire a feel for moonwalking, Bean is strapped into the lunar gravity simulator (**5**), one of many devices used during lunar surface simulation training in Building 29 at the Manned Spacecraft Center in Houston. This simulator approximates lunar gravity by lifting the astronaut with a force equal to five-sixths of his fully-suited weight. (**6**) The crew had to practice even the seemingly-simple routine of picking up rocks. Conrad (left) seems to be telling Bean he wants specimens no larger than he can hold in one hand. On Bean's left hip is a container of plastic bags, in which samples are deposited and sealed, and another roll of bags can be seen on the Lunar Hand Tool Kit in front of the astronauts. This kit includes geological tools and instruments to collect, mark, record, and protect lunar geological samples, and is left on the moon. Seen behind the astronauts is one leg of the L M simulator.

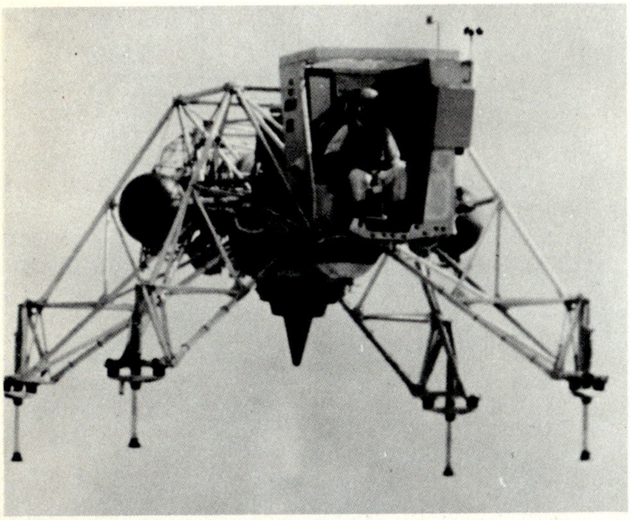

2

4

5

6

APOLLO 12 Pre-Launch

The big day for the nation and the Navy finally arrives—November 14, 1969. Although no subsequent mission could hope to have quite the worldwide attention and historical impact of Apollo 11, this mission would drive home the fact that man's first landing on the moon was not an isolated exploratory incident or an unrepeatable peak of technical achievment, but one of a series of increasingly sophisticated expeditions by which humanity would gradually gain dominion over a new world and become heir to a vast fund of knowledge. Apollo 11 was a supreme adventure, a triumph of the spirit and a demonstration of capability in much the manner of Peary's polar expedition or Lindbergh's flight; Apollo 12 was to use that capability as a tool. (**1**) As the countdown proceeds a few hours before liftoff, a small army of launch personnel monitor Apollo 12's vital functions from Firing Room 2 of the Launch Control Center. Three of the large overhead monitor screens are displaying electrical measurements, which are compared to the

1

3

desired values found on certain pages of the launch sequence program; the monitors are also indicating the proper page numbers. The monitor screen on the right can be switched to any of the numerous closed TV circuits, but is used mainly to follow the astronauts after they have suited up and entered the launch complex. (**2**) Two hours before launch, the sun is brightly shining as the press area of the viewing stands three miles away from pad 39A fills up with journalists from all over the world. Telescopic cameras are much in evidence. Notice that the entire press seating area is equipped with desks and covered with a roof for protection from the Florida sun. But shortly before launch time the sky becomes overcast and it starts to drizzle. This does not dampen the enthusiasm of the crowd, among which is President Nixon (**3**), seen here greeting a well-wisher. This will be the first Apollo flight Mr. Nixon has viewed in person. As were both his immediate predecessors in office, Mr. Nixon is a Navy veteran of World War II, and probably has a special interest in the performance of this mission's all-Navy crew. Typical of all roads leading to the Cape at any launch time, but particularly a lunar flight, is this causeway (**4**), lined with cars on both sides. Many of these spectators have been here for days, living and sleeping in their cars until liftoff. The population of the Cape temporarily doubles during the countdown of a major space shot. Nearby shops, stores and motels enjoy a brisk sale in the carnival-like atmosphere.

2

4

APOLLO 12 Pre-Launch

The crew emerges from the preparation area (**1**) and enters the van which will take them to Pad 39A. Conrad is about to enter the vehicle, Gordon waves at the camera, and Bean smiles. Astronaut Tom Stafford is in civilian clothes at the rear. Up in the white room (**2**), 320 feet above the pad, the crew enters the CM. Conrad (center) is being checked over prior to entry, while Bean waits his turn and chats with the technicians. On the wall above the entry hatch is taped a drawing of the astronauts and a clipper ship, in reference to the CM name *Yankee Clipper.* Note the extra soft caps also taped to the wall. All personnel must wear hard hats while working outside in the vicinity of the rocket, but exchange them for caps upon entering the white room. The Apollo 12 vehicle actually emerged from the Vehicle Assembly Building (**3**) much earlier, on September 8. Seen here as it leaves High Bay 3 of the VAB, the vehicle is carried along with the Launch Umbilical Tower and the Mobile Launch Platform on the huge

1

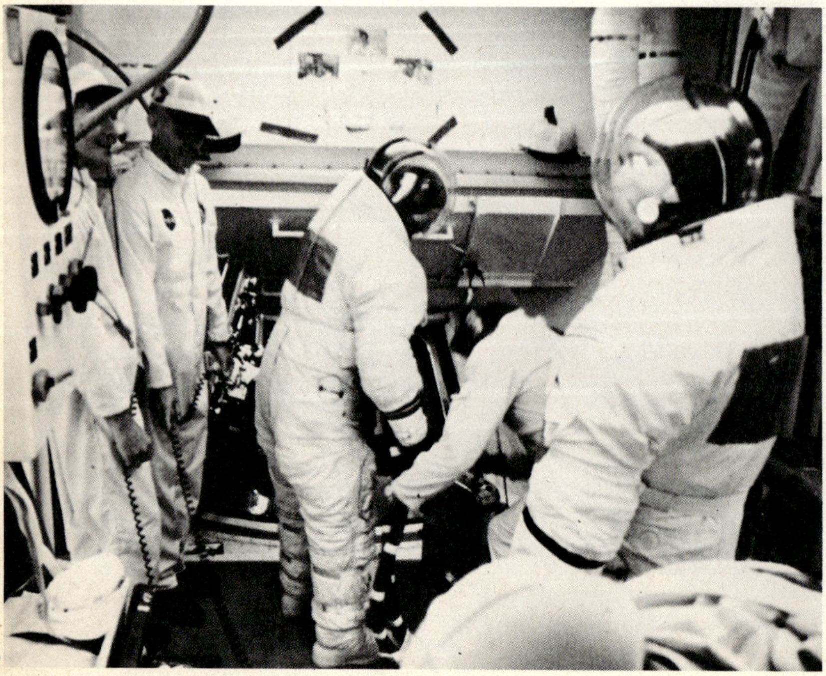

2

3

crawler transporter. Part of the Mobile Launch Platform is in the foreground; the three men on the right are leaning on the platform's railing. (**4**) Viewed from the 525-foot-high roof of the VAB shortly after its rollout, the true 30-story height of Apollo 12 is not apparent. Attached to the upper arm of the tower can be seen the white room, snuggled up against *Yankee Clipper.* Apollo 12 used Saturn V launch vehicle A/S 507, LM 6, and spacecraft 108. As zero hour approaches, Launch Control Center considers the possibility of a hold until the rain clouds clear the Cape, but decides to launch on schedule when a search plane reports no lightning within 20 miles. At 11:22 AM EST Apollo 12 rumbles smoothly to life and rises slowly skyward. Target: the Ocean of Storms. As it enters the low overcast a flicker of lightning flashes between rocket and ground, and instantly the launch becomes anything but routine. From Conrad: "I don't know what happened here. We had everything in the world drop out." The spacecraft's entire fuel cell electrical system automatically disconnected as overload detectors sensed the surge and tripped the circuit breakers. The engines did not shut down since a rocket does not need electricity once ignited, but the inertial guidance system was adrift. The backup battery system takes over the load while Launch Control tells the astronauts what happened. Within three minutes the crew resets the circuit breakers, but the guidance system is not corrected until Gordon takes star sights.

4

APOLLO 12
Launch
_.10
_.09
_.08
_.07
_.06
_.05
_.04
_.03
_.02
_.01
_.00

APOLLO 12
In-flight

Trans-lunar injection for the Apollo 12 mission occurs on schedule, as the crew restarts the S-IVB's J-2 engine to take them out of earth orbit and boost them into the long three-day journey to the moon. During the 83 hours and 25 minutes of trans-lunar flight several television programs are transmitted, showing the crew performing their assigned duties. Due to the location of the landing site, Apollo 12 would fly a different approach path than previous lunar missions. Those had used a "free-return" trajectory, in which the moon would pull the spacecraft around and head it back towards earth in case of engine malfunction; it would be difficult for Apollo 12 to return if trouble developed with the Service Module's SPS engine. (**1**) The interior of *Yankee Clipper* is seen in this reproduction from the third color TV transmission to earth, as Apollo 12 nears the moon. CM pilot Gordon reaches above his head to open the forward tunnel hatch, which allows access to the docked LM. Through this hatch will pass Conrad and

2

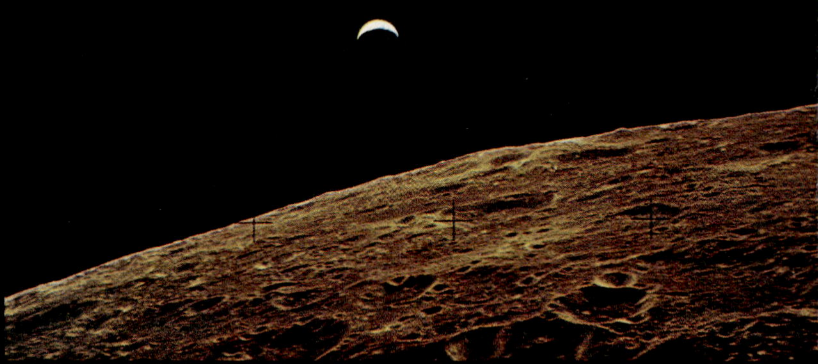

3

Bean when they man *Intrepid*. The approach goes smoothly, and at 10:47:23 PM EST on November 17 the Apollo 12 CSM sweeps into lunar orbit. (**2**) In the center here is crater Lalande, near a proposed future landing site. The large smooth crater in the upper right is Ptolemaeus, and next to it is Herschel; lunar north is to the left in this view. Over the horizon peeps the earth in crescent phase (**3**), a sight all astronauts agree is far more beautiful than the moon seen from earth. (**4**) Still in orbit, Apollo 12 sweeps over an unidentified crater, which the crew televises; this picture was made from that transmission. From *Yankee Clipper*, Gordon records the departure of *Intrepid* (**5**) for the TV audience back home, the first time that an undocking of a moonbound LM was covered live and in color; this event occurs during the fifth TV transmission. (**6**) A small and dear little object in the black immensity of space, *Intrepid* slowly dwindles as it falls behind *Yankee Clipper* and drops down toward the Ocean of Storms. Alan Bean at *Intrepid's* helm has a navigational task that would appall an old-time mariner; in 4½ hours he and Conrad will travel 15,000 miles over forbidding terrain while descending 69 miles. During powered descent Conrad must fly 332 miles in 12 minutes while dropping 50,000 feet, pick out a tiny man-made object (Surveyor 3) in the gray vastness below him, and land his 18-earth-ton craft beside it, all without exhausting the precious fuel supply. Just one mistake would invite disaster.

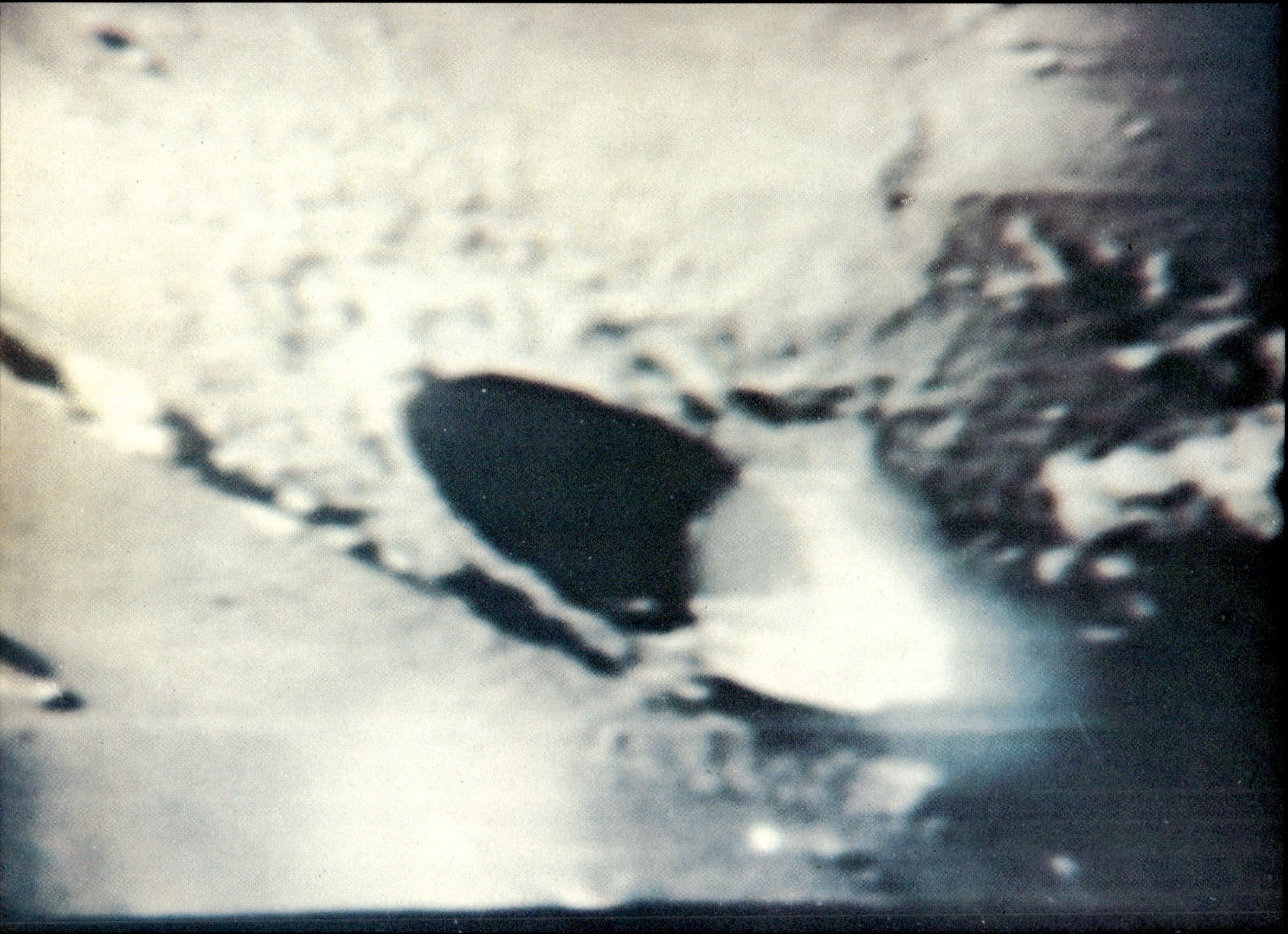

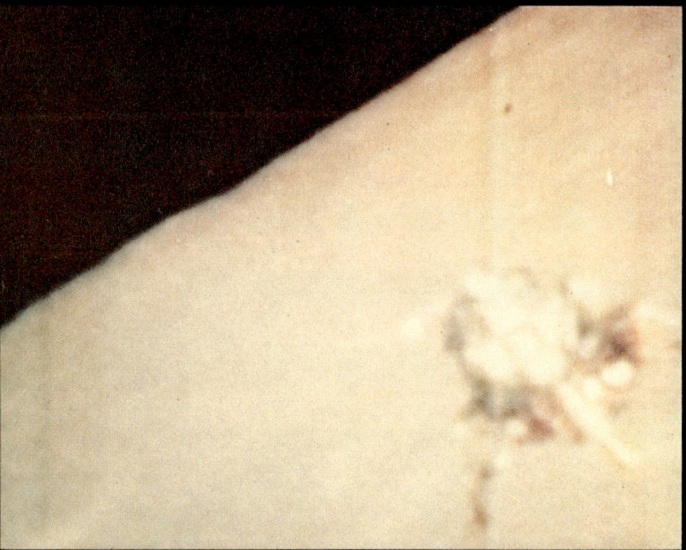

6

Months of practice make perfect, and at 01:54:35 in the morning of November 19 *Intrepid* settles down into the lunar soil, an incredible 25 feet from the rim of the small crater wherein the Surveyor rests. Bean and Conrad can see it clearly, and could throw a lunar pebble to it; this is one of the most pinpoint landings in the history of navigation. Conrad is the first to step out, and he jogs around in high spirits getting used to lunar gravity. "Boy, you sure lean forward," Bean comments. Then he follows, and both astronauts are soon loping around the moonscape with the same peculiar gait used by the Apollo 11 crew. They notice the dust right away, which plagues them even when unloading and setting up the ALSEP components. Conrad remarks to Houston, "There's no way to handle all this equipment with all the dust on it. Every time you move something the dust flies." Because of the weak gravity the dust shoots up quite high when disturbed, but settles rapidly since

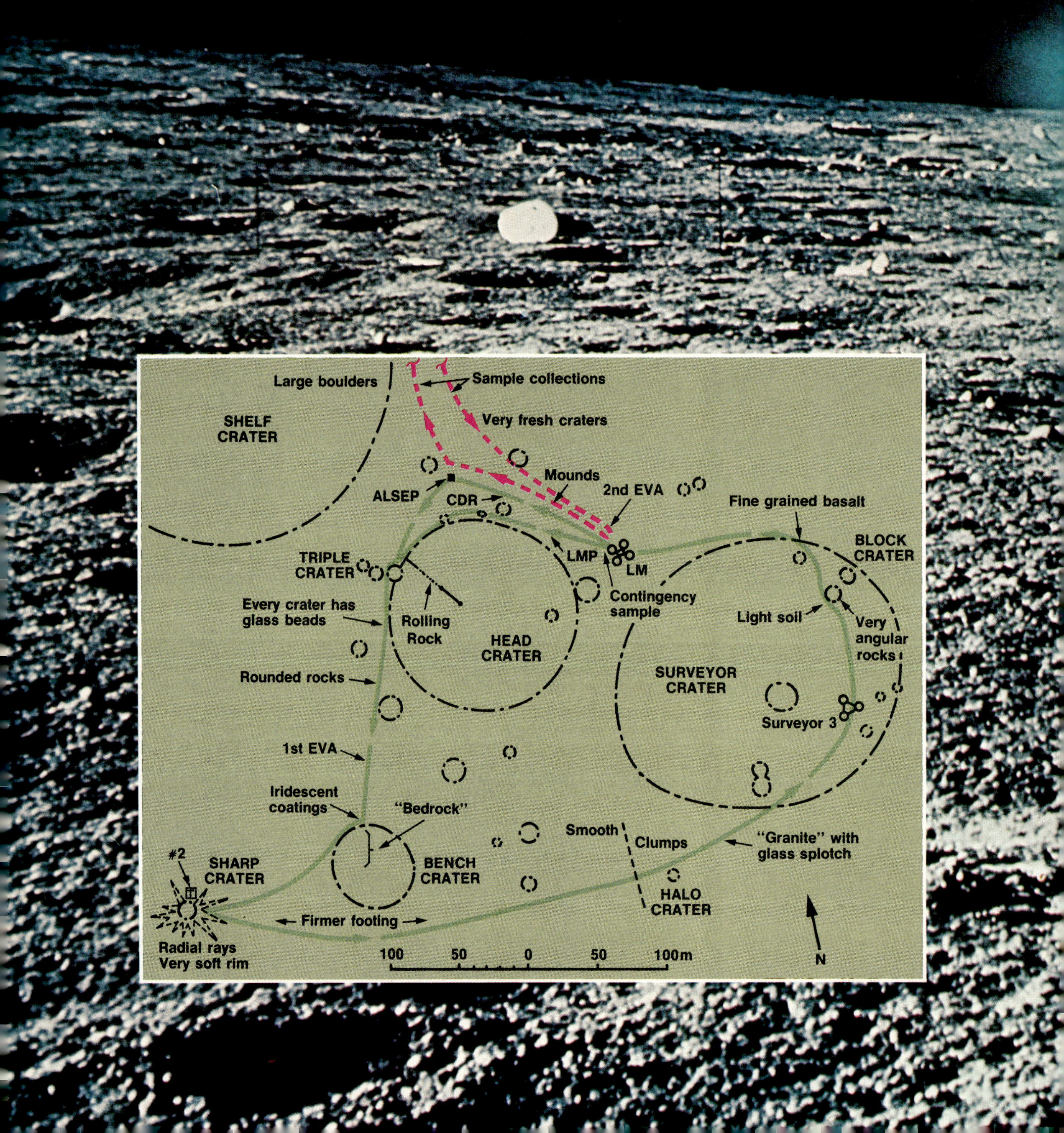

there is no air to support it. Below, sunlight diffracted by the ever-present dust surrounds Bean with a halo as he sets up the ALSEP. Noticed several times, this phenomenon somewhat resembles the legendary St. Elmo's fire of earth, although the causes are completely different. (Inset) This map shows the nearby terrain and the routes taken by Conrad and Bean in their two EVAs. The first EVA, which lasts 3 hours, 56 minutes, is occupied chiefly with the setting up of ALSEP. For this the astronauts select a spot about 600 feet to the northwest of the landing site, so that the instruments will not be disturbed or covered with the dust during liftoff of the LM's ascent stage. Also requiring deployment are the dish-shaped S-band antenna and the solar wind experiment, neither of which is part of ALSEP. When Bean removes the plutonium fuel rod from its protective sheath and inserts it into the SNAP-27 generator the ALSEP experiments begin operating, and the crew is free to become lunar rockhounds. They take a short trip to the north to explore the largest nearby crater, then return to *Intrepid,* taking photos and picking up samples along the way. One of these Conrad describes as "a pure piece of glass." After 5 hours of sleep and a conference with Houston about the proposed route for the next EVA, the astronauts embark on a one-mile trip that brings them back to the LM 3 hours and 49 minutes later, loaded with moon rocks.

APOLLO 12
Exploration

(**1**) A good view of the almost alien appearance of *Intrepid* in the Ocean of Storms. If Lunar Modules were much bigger, they would look like those Martian invasion machines from the movie version of H.G. Wells' novel "War of Worlds." (**2**) The S-band antenna next to *Intrepid,* as one of the astronauts loads the ALSEP experiments on the bar-bell for transportation to the set-up site. This antenna is the first device the astronauts erect, for it augments their communication signal strength to earth. It greatly resembles an upsidedown umbrella, and indeed unfolds exactly like one. On the airless, waterless moon the sun glares down with a pitiless intensity unequalled anywhere on earth. Bean's camera magnifies this effect (**3**), but the suggestion of truth remains as Conrad is seen trudging along with the ALSEP bar-bell. This is the color TV camera (**4**) that unfortunately ceased operating shortly after it was set up. Although it was carefully placed with the sun behind it, the camera is believed to have been blinded by

1

2

5

sunstroke—a strong solar reflection that overloaded a vital circuit. (**5**) Bean carefully removes the SNAP-27's plutonium fuel rod, encased in its protective shield, from its storage place aboard *Intrepid*. The electrical generator in which the rod will be inserted to provide power for ALSEP is in the foreground. At the ALSEP site (**6**) the astronauts set up the instruments. In the foreground in its folded configuration is the surface magnetometer, designed to measure the moon's magnetic field. During the second EVA both astronauts comment that every crater examined contains a number of glass beads. Near the small crater called Sharp, distinguished by its white rim and radial pattern of rays, the team notes that the surface is much softer than elsewhere. They take core samples and move next to an even smaller crater named Halo, where the surface is harder and has many clumps. Here they have to use the coring tube's hammer to force the tube down to its full length. Conrad notes to Houston that so far they have not encountered any breccia, the mineral-laden type of rock brought back by Apollo 11. They then proceed to their great prize, Surveyor, (**7**) only 31 months old yet already a nearly-forgotten relic from an earlier era of space exploration. They examine and photograph the robot probe in detail before removing its trenching scoop, camera, several pieces of tubing, and a cable. Despite its lengthy exposure, the astronauts agree that Surveyor is in excellent condition and could have been made to operate again.

4

7

APOLLO 12
In-flight

Finally the astronauts return to *Intrepid* and clamber aboard with their samples, which total 75 pounds for the two EVAs. After a long checkout of the ascent stage, the lunar team announces it is ready for liftoff. After 31 hours and 31 minutes on the moon, Bean and Conrad ignite *Intrepid's* ascent stage engine at 9:25:47 AM EST in the morning of November 20, and the two stages separate violently as the astronauts climb toward a matching lunar orbit with Gordon. Seven minutes and five seconds later the constant-thrust engine has pushed *Intrepid* into orbit at a speed relative to the moon of slightly more than 5000 feet per second, but still more than 140 miles from *Yankee Clipper*. The two craft slowly maneuver toward each other, and after several hours gain visual contact at less than a mile. Gordon greets his returning partners with a breezy "Hey Pete! How can you look so good if you're so ugly?" Shortly afterwards they are station keeping, and Bean rotates *Intrepid* in all directions so that Gordon can

1

inspect its fragile skin for any damage. Then begins the cautious final approach, as Gordon inches *Clipper* toward the docking. He finally announces, "You're home free, boys," and the *Intrepid* pair pay him an ultimate compliment by noting that they didn't even feel a ripple. They then transfer aboard the CM, and the crew is together again. *Intrepid* is undocked but kept close by, for its work is not yet finished. Its engines are activated by remote control, and at 5000 miles per hour the ascent stage crashes into the moon 45 miles away from its previous landing site. Not only does the known force of this blow provide a reference point for other tremors registered by the two seismometers now on the lunar surface, it reveals the astounding fact that the moon's internal composition propagates shock waves in a manner totally different from earth's. The waves intensified to a peak 8 minutes after the crash, then diminished to end after 55 minutes; on earth the whole performance would have been over in two minutes. In one geologist's words, "The moon rang like a bell." *Yankee Clipper* leaves lunar orbit at 3:49:16 EST on November 21. On the way home Bean enjoys the blissful luxury of a shave (**1**) while Conrad watches critically. A highlight of the return was the solar eclipse by the earth (**2**), an event never before witnessed. During their ninth TV transmission (**3**) the astronauts held a long-distance press conference, answering questions about the historic flight from newsmen assembled in Houston.

2

3

APOLLO 12 Splashdown

After travelling almost a million miles during its round trip, *Yankee Clipper* smacks down hard (**1**) in choppy seas 350 nautical miles southeast of Samoa at 3:58:24 PM EST on November 24. The spacecraft is right on target, impacting only three miles from that familiar prime recovery ship, the carrier USS *Hornet.* The total elapsed time of the Apollo 12 mission, from liftoff to splashdown, was 244 hours, 36 minutes, 25 seconds. Wearing their germicidal masks and flight coveralls (**2**), Gordon and Bean steady Conrad as he climbs into a life raft while a Navy pararescueman closes the hatch. Like a good Navy skipper, mission commander Conrad is the last to leave his craft. Note how much of *Clipper's* surface coating peeled off from the intense heat of re-entry. (**3**) The helicopter's rotor blast smooths the waves as the astronauts are hoisted aboard. On the *Hornet,* (**4**), the astronauts wave to the assembled crew as they are led immediately to the isolation van, or Mobile Quarantine Facility. They have already donned the baseball caps

1

2

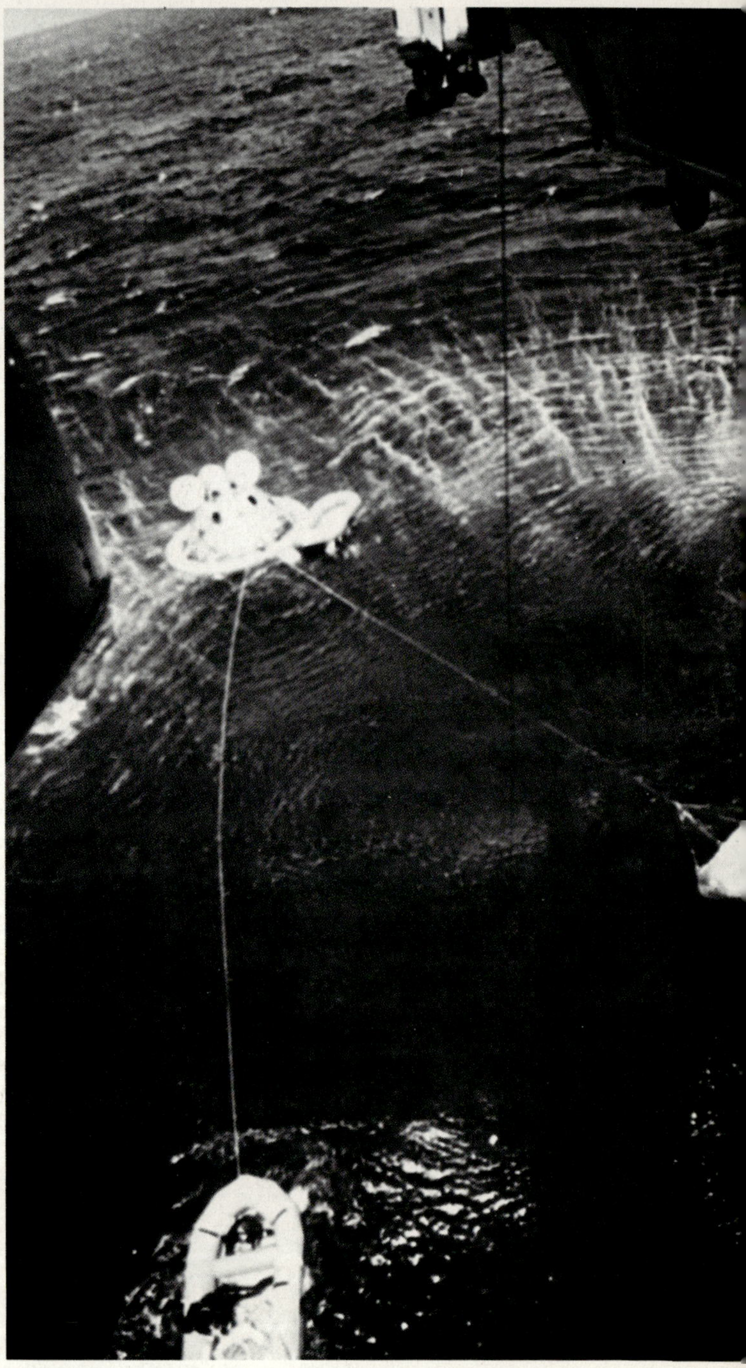

3

that the Navy traditionally gives to returning astronauts. Above the door of the patriotically-striped van the slightly-ungramamatical sign reads "Three More Like Before", a reference to the *Hornet's* recovery of the Apollo 11 team. (**5**) The ship's chaplain offers a thanksgiving prayer during welcoming ceremonies, which the astronauts view from their quarantine van. As the banner indicates, the recovery operation nearly coincides with the *Hornet's* 26th birthday—it was commissioned on November 29, 1943, the successor to that famous carrier of the same name which lauched Doolittle's Tokyo raiders—so the ship's crew celebrate the two events together. (**6**) Scorched but intact, *Yankee Clipper* is hauled close to the *Hornet* prior to being hoisted aboard. (**7**) This is a close-up view of one of Apollo 12's core samples. Labelled sample number 12025, it was obtained at a point about 225 meters from *Intrepid's* landing site. All core samples from Apollo 12 differ considerably from those obtained in the Sea of Tranquillity by the Apollo 11 crew. The new samples all have easily recognizable stratigraphy and two coherent crust-like layers; this particular sample has a predominately fine-grained structure, somewhat like riverbed silt on earth. This dazzling jumble of shades (**8**) is a microphoto taken with polarized light of a thin section. It is officially designated sample number 12063,6, meaning the 6th slice from the 63rd sample of the Apollo 12 mission. Supplying a feast for scientists, Apollo 12 indeed provided a new look at the moon.

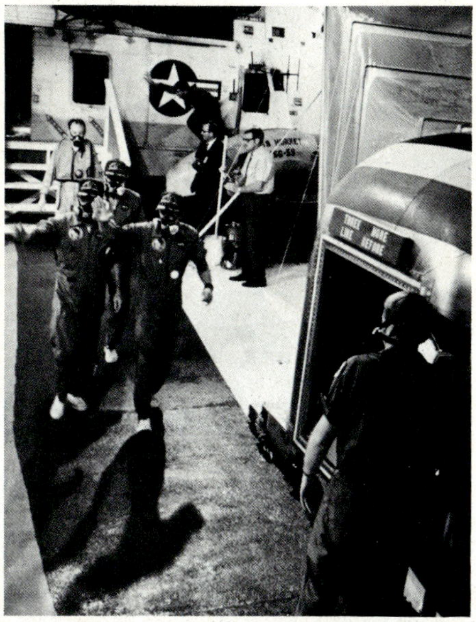

4

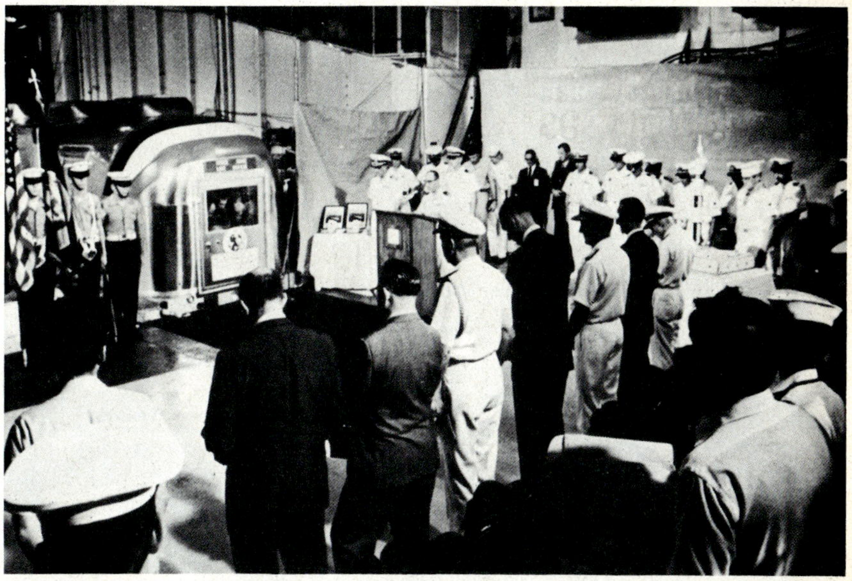

5

6

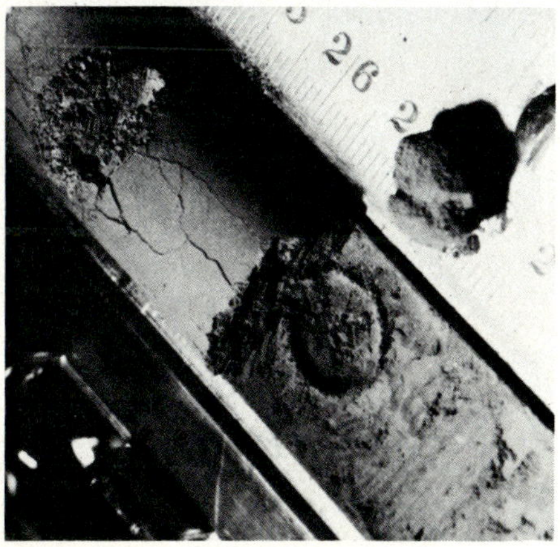

7

8

Building 37 at the Manned Spacecraft Center in Houston is more commonly known by its formal title, the Lunar Receiving Laboratory. It is a self-contained enclave capable of isolation, examination, and decontamination of all material from the moon, including the command module and the astronauts themselves. (**1**) July 25, 1969—an historic event in the history of exploration, as the first lunar rocks from the Apollo 11 mission are unloaded at the LRL only a few hours after splashdown in the Pacific. The first material from another planetary body, these samples arrived at the LRL before the astronauts. (**2**) Part of the LRL is the Sample Operations Area, shown here in cutaway. On the center floor is the Primary Vacuum Laboratory, containing chambers for handling and examining samples under vacuum conditions. Plumbing connects these processing chambers with mass spectrometers and other analytical and recording equipment in the area on the lower left, while on the top floor are controls, comput-

THE LUNAR RECEIVING LAB

NASA's methods of preventing contamination from space

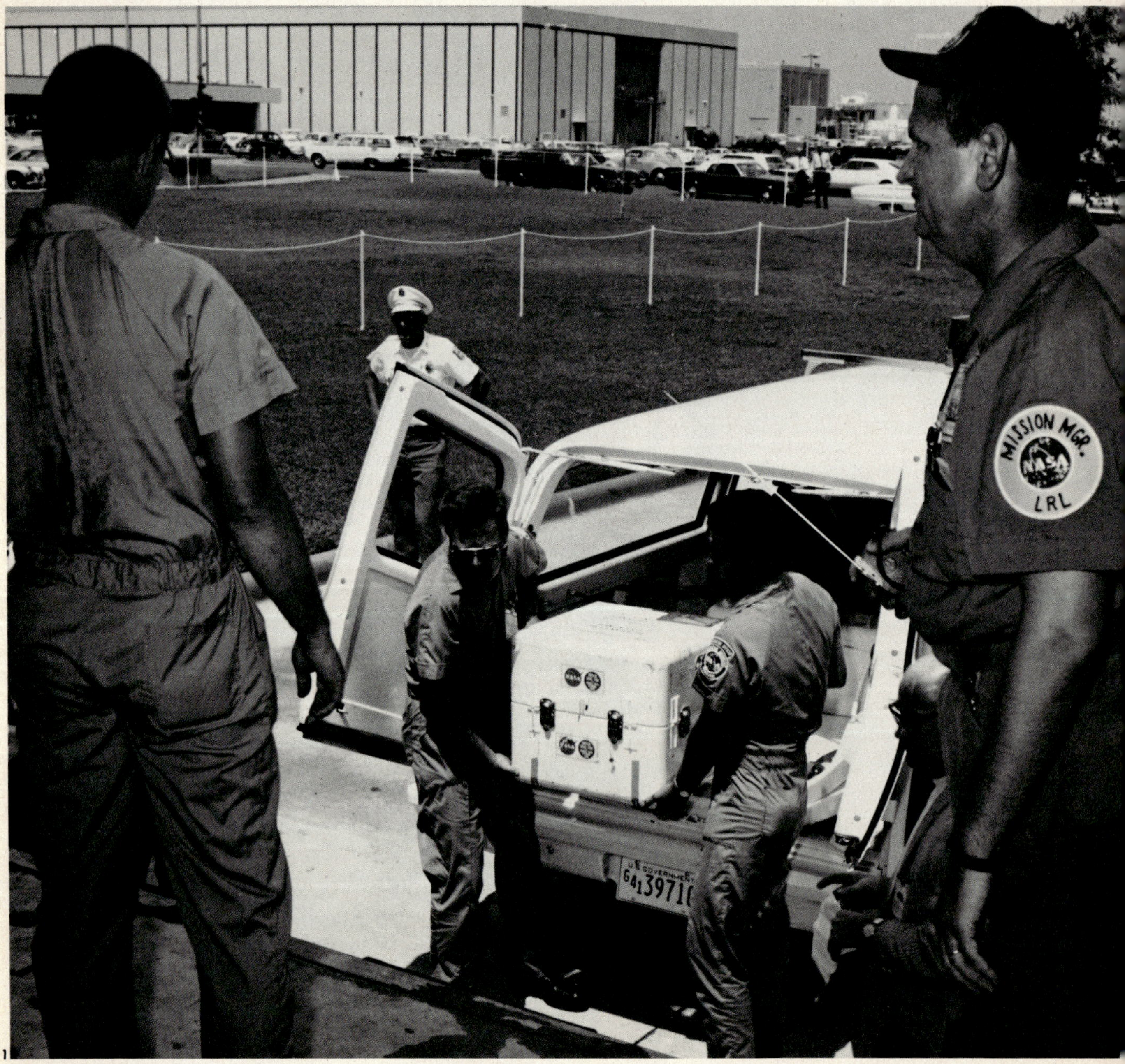

1

ers, and electronic storage units. Accessible only by elevator, the Radiation Counting Laboratory (**3**) is shielded by concrete and buried in the ground beneath administrative offices and lab rooms. Here lunar samples are examined for radioactivity in the area surrounded by dunite, which is a rock shield containing very low amounts of potassium, thorium, and uranium. These elements have natural radioactivity and would cause counting interference. The anti-coincidence shield prevents counting confusion between any inherent gamma radiation in the sample and secondary radiation caused by atmospheric cosmic ray bombardment. Transportation of lunar materials to the LRL is accomplished in several ways (**4**). The geological samples, all recorded data, and biological specimens are the first to arrive, being flown directly to Houston from the recovery ship. The astronauts and command module are carried by the ship to the nearest land base with large runways to accommodate heavy transport aircraft, as the astronauts' isolation van is too bulky to be airlifted by carrier-based aircraft. The spacecraft is then brought to Houston separately. The LRL is not capable of performing all desired tests, so NASA seeks the assistance of other laboratories. Part of the LRL is the Crew Reception Area where the returning astronauts are quarantined for twenty-one days. Apollo 11 CM pilot Mike Collins (**5**) samples the motel-like surroundings where the crew remains until any possible extraterrestrial micro-organisms present have been isolated and neutralized.

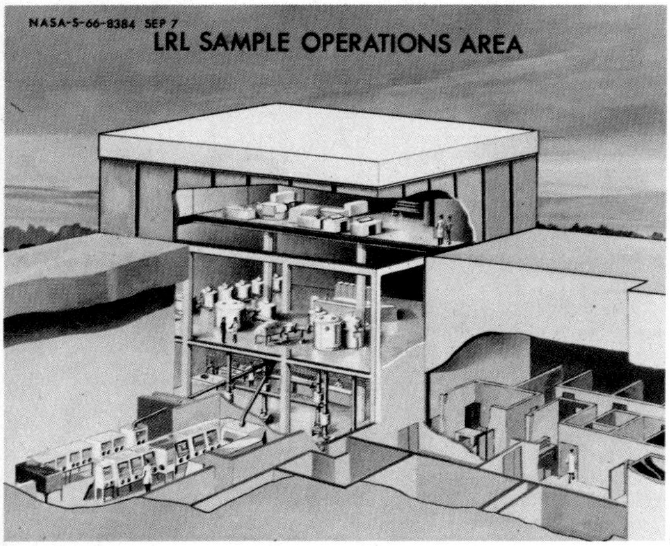

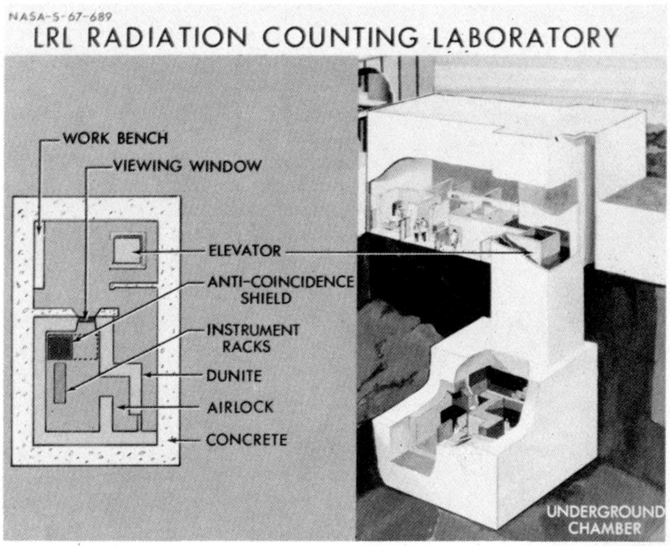

3

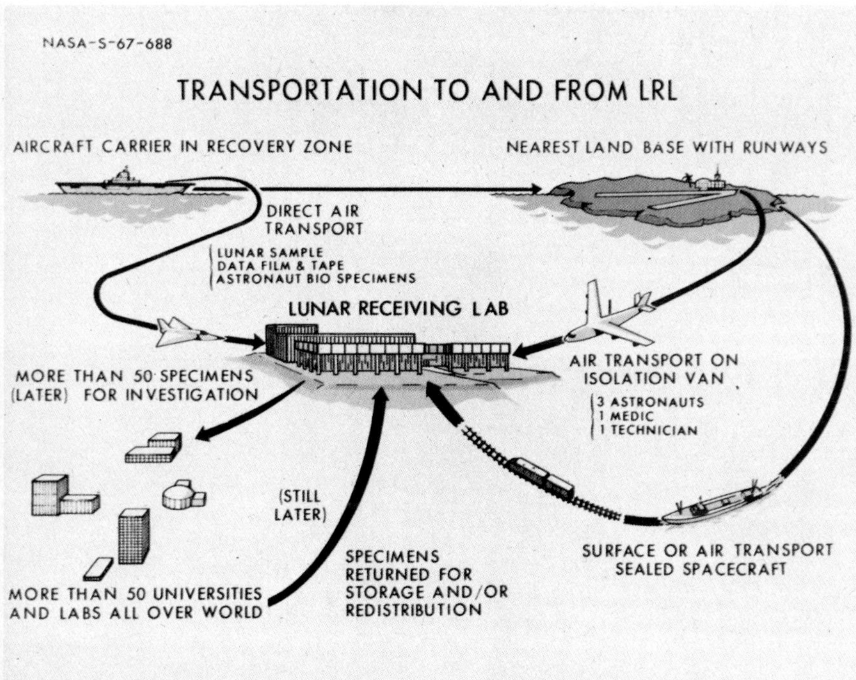

5

THE LUNAR RECEIVING LAB

Arriving lunar samples must first be decontaminated from contact with the earth's atmosphere. This is done in the Atmospheric Decontamination System (**1**), located in the Primary Vacuum Lab. The sealed box containing samples is placed in the first compartment on the right, numbered R-101, where it is vacuum cleaned. It is then transferred through an internal door to compartment R-102 and washed with distilled water, then dried with nitrogen in compartment R-103. Still unopened, the box is then transferred into the primary vacuum system for further processing. Control operators can handle the box in each compartment with externally-sealed rubber gloves. The entire system is maintained at a slight vacuum so that any micro-organism cannot escape into the atmosphere. The box is opened in Carousel F-201 (**2**), a vacuum chamber so named because it contains revolving trays for samples and tools. Also inside the Carousel are weighing scales, a photography platform, containers for sealing individual

2

1

samples, and tools for obtaining chips for chemical analysis. The operator stands in front of the chamber and reaches in with sealed gloves (not shown) through the two lower holes. The upper hole is the viewport, also used for photography. In the Thin Section Laboratory (**3**), a wire saw is used to prepare thin slices of material for crystallographic photography and chemical treatment. In the Biological Test Laboratory (**4**), executive assistant Richard A. Wright (left) explains the operation of one of the autoclaves to astronaut Mike Collins. Applying both heat and pressure, these chambers are used primarily for preparation and sterilization of tools and equipment. At the Invertebrate, Aves, and Fish Laboratory (**5**), Wright and Collins examine the vacuum cabinets used as holding and observation areas for various creatures exposed to lunar material. For fish and birds, small amounts of the material are placed in their food; direct injections are administered to the invertebrates. A similar laboratory conducts exposure experiments on mammals and primates. The results from these labs, plus those from the medical facilities, were responsible for the initial establishment of the 21-day isolation period for returning astronauts. During this time the crew is completely isolated. They communicate with the outside by microphone, and eat food already in the LRL. NASA takes no chance that a biological disaster such as that postulated by author Michael Crichton in his best-selling novel "The Andromeda Strain" could ever occur.

5

THE LUNAR RECEIVING LAB

The Biomedical Laboratory of the LRL (**1**) is a joint operation by scientific and medical personnel. Here are performed routine and contingency clinical chemistry tests, immunology, and hematology evaluations upon astronauts and support personnel. Major instrumentation includes spectrophotometers capable of operating in three light frequency ranges—ultra-violet, visible, and infrared—as well as the atomic absorption type. Also available is an analytical ultracentrifuge, a Coulter counter-multichannel analyzer, gas chromatograph, amino acid analyzer, Beta scintillation counter, automatic gamma radiation counter, several auto-analyzers, and a robot chemist. For instant data storage and quick reference later, the major instruments are directly connected to a computer located in another section of the LRL. Also in the Biomedical Lab are typically found such minor but high-quality instruments as the Spencer stereo microscope (**2**), used for slide analysis as shown here. Atop the microscope is a large plate camera. (**3**)

1

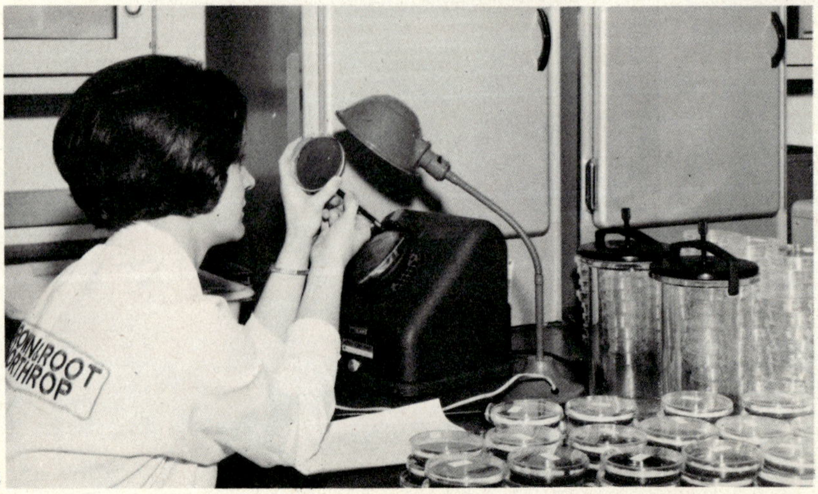

4

5

Among the several rooms of the Spectrographic Laboratory is the spectrographic dark room, in which is located this Jarrell-Ash spectrographic plate and film photoprocessor. Room 137 of the LRL is the Microbiology Lab (**4**), where this inspection of an anaerobically incubated culture is taking place. This lab is used for food and water microbiology, germ-free animal microbiological monitoring, and quality testing of material to be used in the Sample Laboratory. Equipment used here includes light and stereo microscopes, high and low temperature incubators, microaerophilic incubating apparatus, automatic colony counters, and vacuum ovens. (**5**) Here is the gamma ray detector being withdrawn from its inner anticoincidence mantle in the Radiation Counting Laboratory. The rectangular structure surrounding the mantle is the outer anticoincidence shield, on which a lid fits. Properly isolated from cosmic ray interference by this double-layered protection, the sensitive detector can locate the tiniest amount of radioactivity in a lunar sample. In the Gas Analysis Lab (**6**), operator David R. Moore vaporizes a small amount of lunar material by radiofrequency induction heating. The gas is then analyzed to determine its constituents. As with all NASA work, nothing in the LRL is secret; the test results from its many laboratories are disseminated widely and available to scientists everywhere. As impressive as Cape Kennedy's launch facilities are, the LRL becomes the most important place in NASA after astronauts return.

3

APOLLO 13

Four hundred miles south of the barren plains of the moon's Sea of Showers (Mare Imbrium), across the dusty sweep of the Seething Bay (Sinus Aestuum), lies a hilly region of great interest to selenologists. The crater Fra Mauro (named for a cartographer-monk of the Italian Renaissance) and its environs appear to be ancient craters that were inundated eons ago when a gigantic meteor struck the moon and formed the Mare Imbrium, which itself resembles an enormous crater. When did this happen? Did the moon then possess a molten core which flowed out, or was the force of this titanic impact sufficient to melt a vast area? Scientists hope to find at least part of the answers to these and other questions by sending men to explore first-hand this remote region of desolation and ancient ruin, the mission of Apollo 13. Led by the experienced Jim Lovell, the crew named their Command Module *Odyssey* after Homer's epic of the wandering Odysseus, and their LM is christened *Aquarius,* the title of a popular song that portrays the earth moving into a new age. The mission emblem shows three horses pulling the sun god Apollo's chariot from the earth to the moon; the Latin phrase translates, "From the moon, knowledge."

JAMES A. LOVELL, JR.

SPACECRAFT COMMANDER/ With three previous space flights including a lunar mission, genial Navy Captain Lovell is NASA's most experienced and most-decorated astronaut. He was the pilot on Gemini 7, command pilot on Gemini 12, and CM pilot for the maiden voyage to the moon, Apollo 8. Much in demand, he was also twice a backup Gemini pilot and backed up Neil Armstrong on Apollo 11; obviously NASA considers him a pilot's pilot. An exercise buff, he is a President's Consultant on sports and physical fitness.

T. KENNETH MATTINGLY, JR.

COMMAND MODULE PILOT/ Born in Chicago, Tom Mattingly spent most of his life in Florida, and received a BS degree in Aeronautical Engineering from Auburn before joining the Navy in 1958. Now a Commander, he had five years of fleet carrier service as a jet pilot, and was in the Air Force Aerospace Research Pilot School when tapped by NASA in 1966. A part of the support crews for Apollo 8 and 11, he was the astronaut representative in developing the Apollo space suit and backpack. He enjoys water-skiing and tennis.

FRED W. HAISE, JR.

LUNAR MODULE PILOT/ Born 36 years ago in Biloxi, Miss., Haise had a remarkably varied career before being selected as an astronaut in 1966. Originally a Navy flight instructor, he was in succession a Marine fighter pilot, an Air National Guard interceptor pilot, and an Air Force fighter pilot. He then became a civilian research pilot at NASA facilities in Cleveland and Edwards, Calif. He has 7000 hours flight time, four children, and holds a BS degree in Aeronautical Engineering from the University of Oklahoma.

APOLLO XIII
EX LUNA, SCIENTIA

APOLLO 13 Preparation

Practically before the Apollo 12 team returned safely from their expedition, the crew of Apollo 13 is already in serious training for their flight. In mid-January LM pilot Fred Haise (**1**) takes a make-believe lunar stroll in Building 29 at the Manned Spacecraft Center in Houston, assisted by the lunar gravity simulator which relieves him of five-sixths of his weight. Although twice a backup LM pilot, Haise has never been in space before, and has been an astronaut only four years; but he is so well liked and respected by his fellow astronauts, particularly Jim Lovell, that on the 1968 Apollo 8 mission the crew named a crater on the back side of the moon after him. In the same general area are also craters named for the deceased crew of Apollo 1. Space food has come a long way since the toothpaste-tube days of Project Mercury. Here (**2**) being prepared are some bologna sandwiches with whole wheat bread, similar to what will be used during the mission. This young lady with the rubber gloves is an employee of Technology

1

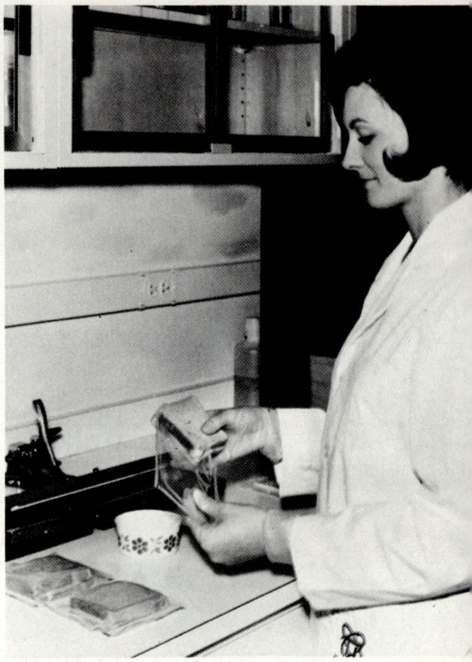

2

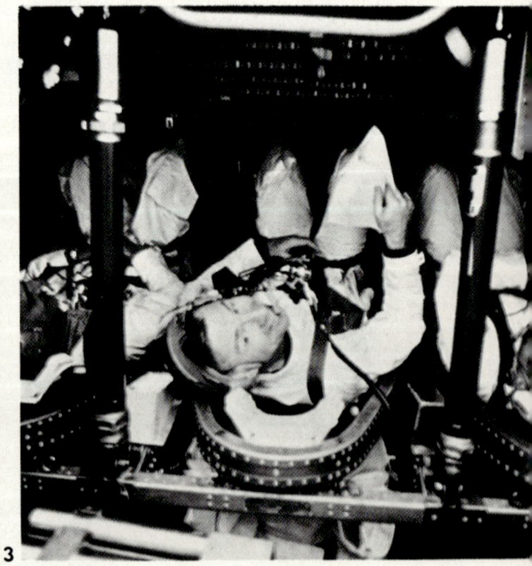

3

Incorporated, which is contracted to provide some of the food; she is probably smiling over the thought that astronauts eat bologna sandwiches like everybody else. Lovell smiles confidently out from the center seat of a special Command Module (**3**) as the crew prepares for water egress tests in February. He is flanked by CM pilot Tom Mattingly (left) and LM pilot Haise. This view is through the opened hatch of the CM, with the spacecraft's heat shield down; the astronauts are flat on their backs. To exit, they must reach up, grab the handle visible at the top of the photo, and pull themselves out backwards, one at a time. The test shown here is a special one for contingency conditions, for normally Lovell would occupy the spacecraft commander's seat at left, and Mattingly the CM pilot's seat in the center. The exiting sequence is CM pilot first, followed by LM pilot and then the Commander. (**4**) March 25, 1970 - it's Jim Lovell's 42nd birthday, and the old pro doles out a piece of the cake presented to him by his co-workers in the Flight Crew Training Building at Kennedy Space Center. On the floor behind Lovell is an enormous birthday card with the signatures of practically everybody in Florida. Another view of the card (**5**) shows its size, with George Killmer of KSC's Public Affairs Office pointing out some signatures to Lovell. Tom Mattingly had a similar celebration for his 34th birthday the week before. During a practice moonwalk on the beach at Cape Kennedy (**6**), Lovell and Haise assemble the lunar drill with lunar hand tools.

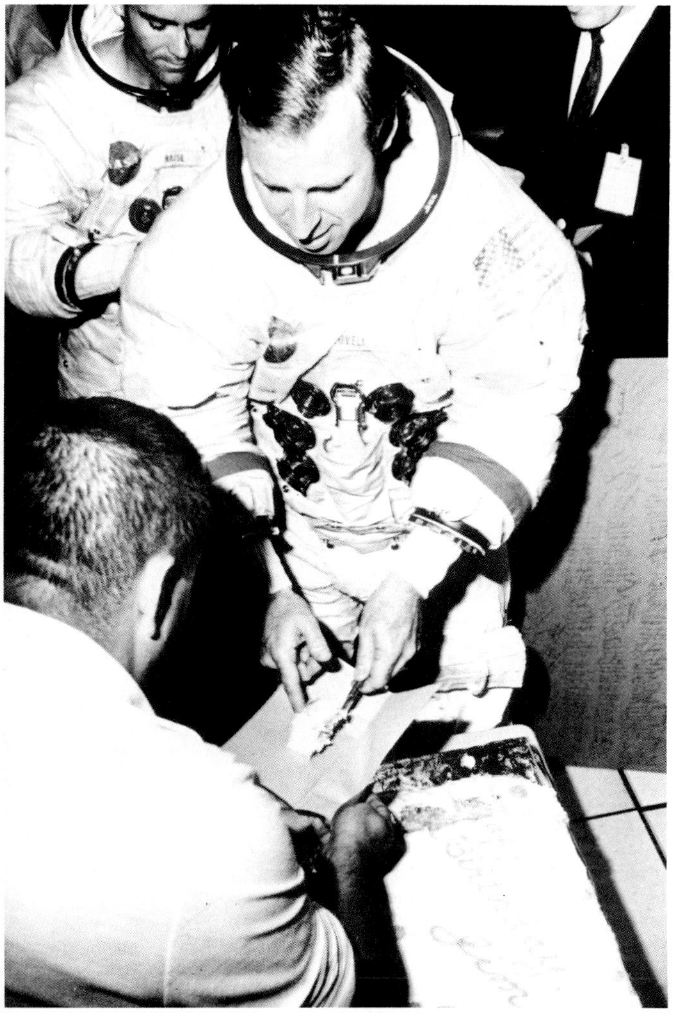

6

APOLLO 13
Pre-Launch

A close-up of Haise (**1**) shows him in a typical grin before another practice moonwalking session. Haise again (**2**) inside the LM simulator, wearing the full space suit without gloves or helmet. (**3**) In early December the Command/Service Module #109 is moved to the integrated work stand at KSC's Manned Spacecraft Operation Building for final mate to its adapter structure and skirt. The trademark on the garments identifies the workers as employees of North American Rockwell, builders of the CSM. Wrapped in a protective shroud (**4**), the CSM with its adapter is brought into the VAB for eventual mating with the launch vehicle, A/S 508. Having passed the vital altitude chamber tests, the descent and ascent stages (**5,6**) of Apollo 13's Lunar Module No. 7 are brought into the low bay work area of the Operations Building. The nozzle of the ascent stage's constant-thrust Rocketdyne engine is clearly seen. (**7**) In the early dawn of a working day at the Cape, floodlights illuminate the CSM, now resplendent in its coating of white ther-

1

2

3

5

6

mal paint. The launch escape system has been added, and the adapter skirt is visible as the assembly is brought into position to be placed atop the Saturn stack. December 15, 1969 (**8**); rollout day for Apollo 13. From a camera high on the Mobile Service Structure, the completed Saturn V launch vehicle A/S 508 is seen as it, the Launch Umbilical Tower, and the Mobile Launch Platform are carried by the crawler transporter out to Pad 39A, visible on the left in the distance. Only 72 hours before scheduled liftoff on April 11, a major snag develops that threatens postponement of the entire mission. Despite medical precautions, backup LM pilot Charles Duke develops a case of rubella (German measles). Since he has worked closely with the prime crew, they are carefully checked. Lovell and Haise are found to be immune to rubella, but not Tom Mattingly; the doctors ground him. The normal procedure would be to bring in the entire backup crew; but obviously Duke also is out of this mission. Deke Slayton proposes to substitute only backup CM pilot Jack Swigert, as he is also found to be immune to rubella. The final decision is left to Jim Lovell, and he balks; he does not care to have, at the last minute, a new member with whose knowledge, reactions, and temperament he is unfamiliar. With the countdown on the rocket proceeding, for two days Swigert trains intensively with the prime crew; here (**9**) he reviews flight procedures with (from left) Slayton, Lovell, and Mattingly. With 24 hours left, Lovell relents, and the flight is on.

8

9

APOLLO 13
Launch

Despite the difficulties of the previous few days, at breakfast on launch day (**1**) Slayton can still find something amusing to say to the Apollo 13 crew seated across from him. Backup LM pilot Charles Duke is seated at far left, slowly recovering from his bout with German measles. The astronauts suit up in the Operations Room (**2**), attended by technicians. From foreground to rear: Lovell, Swigert, and Haise. A craning crowd greets the team (**3**) as they emerge from their suiting preparations at the Manned Spacecraft Operations Building and enter the van which will take them eight miles to Pad 39A. (**4**) ''End of the line, everybody out!'' The van driver gets his chance to jokingly boss the astronauts as they disembark here, where there are no crowds in the shadow of the Launch Umbilical Tower. From here they proceed up to the white room, where they are given a final check before entering the CM *Aquarius*. A victim of the last-minute crew switch (**5**): the crew plaque attached to the Lunar Module descent stage still

3

1

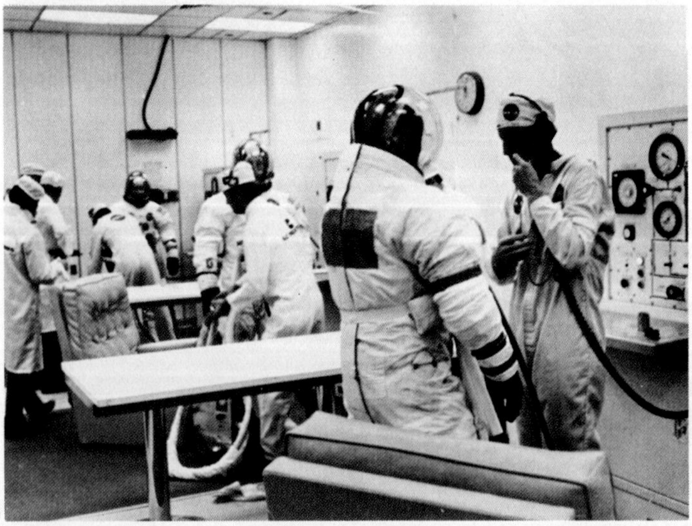

2

4

reads "Thomas K. Mattingly" quite clearly, and there is no time to change it. Up in the Launch Control Center, there are distinguished visitors (**6**). Dr. Kurt Debus, Director of the Kennedy Space Center, presents a spaceport hardhat to His Excellency Willy Brandt, Chancellor of the Federal Republic of Germany, who is visiting the United States with his wife. Accompanying Mr. Brandt from Washington are Vice-President Spiro Agnew (between Debus and Brandt) and NASA Administrator Dr. Thomas Paine (right). Mr. Brandt also received the model of the Apollo Saturn rocket at left. The target area, alternate landing sites, and projected EVA treks of Apollo 13 are depicted here (**7**). The countdown proceeds smoothly, and at 2:13 PM EST on April 11, 1970, the liftoff produces the spectacle of contrasts on the following two pages: the mighty Saturn V soaring heavenward, its white pillar of fire reflected in the still waters of a tree-lined lagoon inhabited by tranquil ducks. It is impossible to capture the true atmosphere of an Apollo launch, for mere pictures cannot convey the intense emotions that overcome all spectators at liftoff—it is a near-religious experience to actually witness three human beings soaring aloft astride a controlled fury so powerful it could become the largest non-nuclear man-made explosion in history should this frail contrivance fail to contain it. Nor is the sound a mere rumble; from a distance the voice of Apollo has a peculiar cracking quality, as if it were rending asunder the very structure of space itself.

6

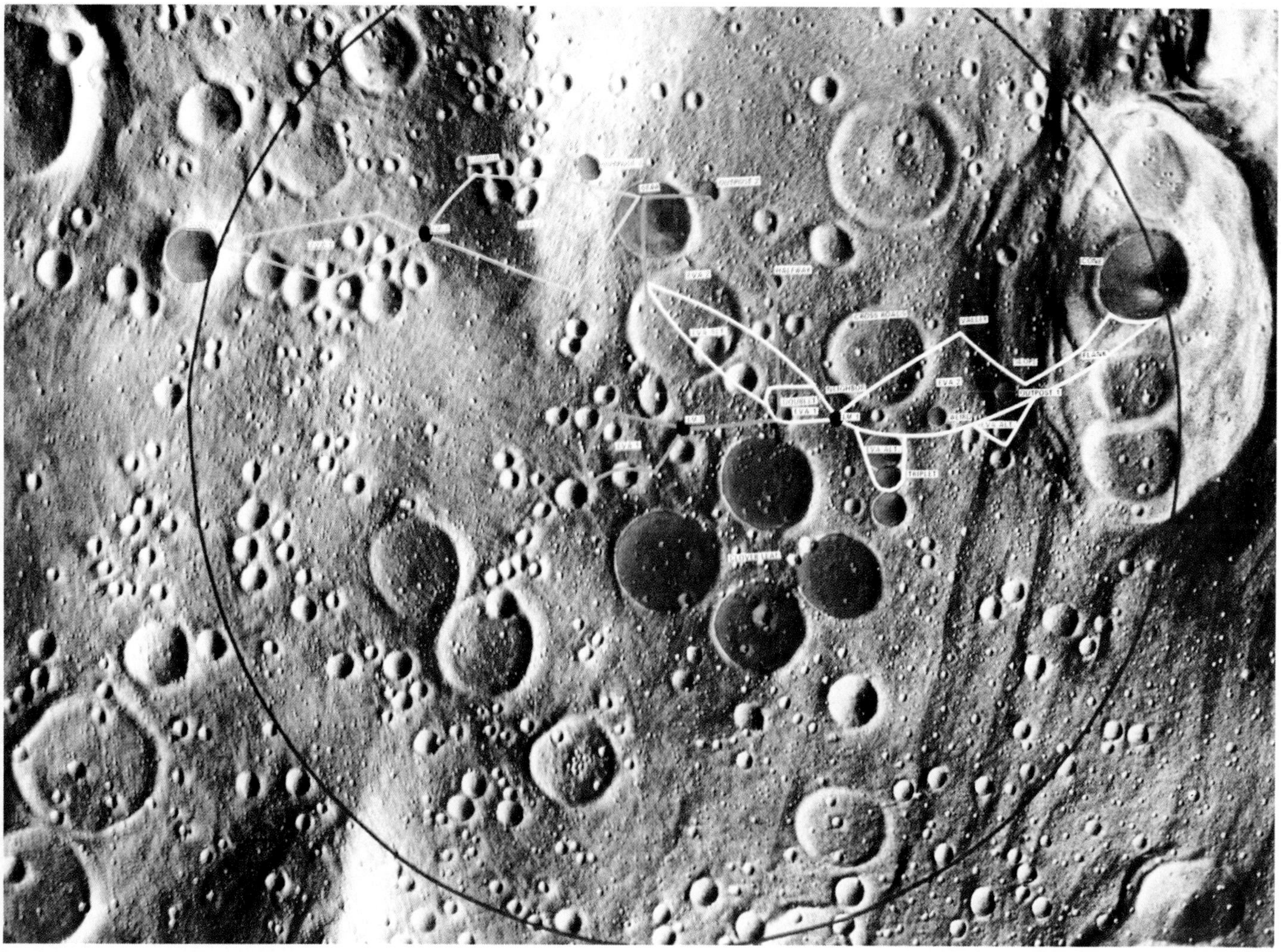

APOLLO 13
Launch

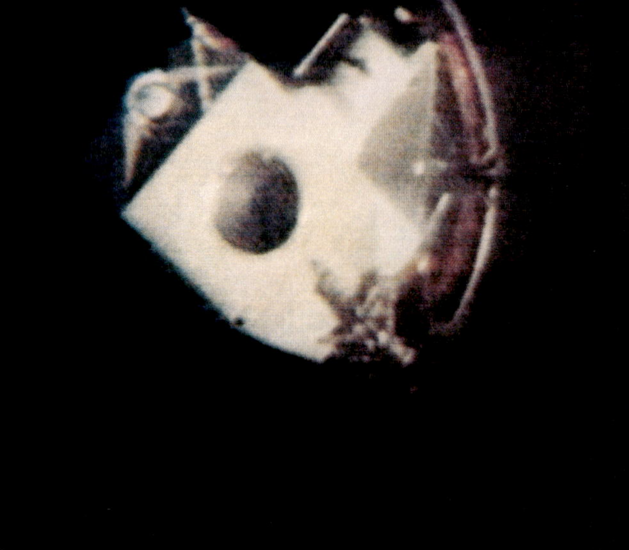

1

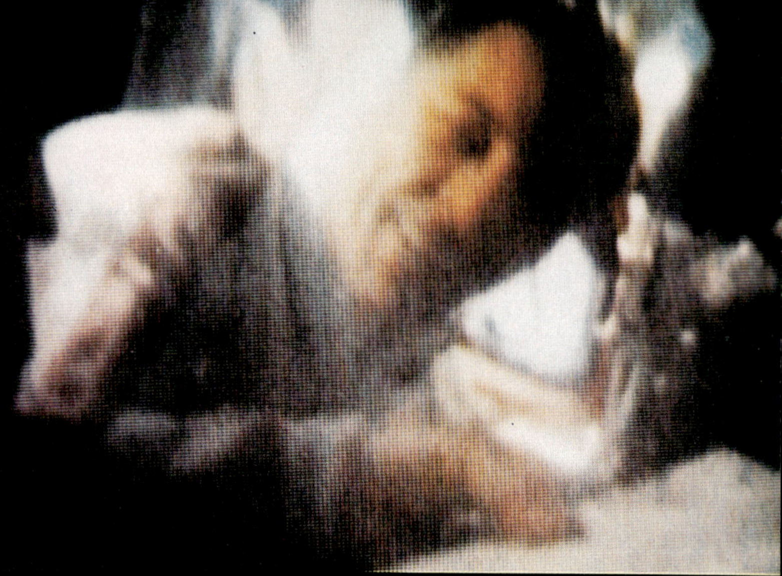

2

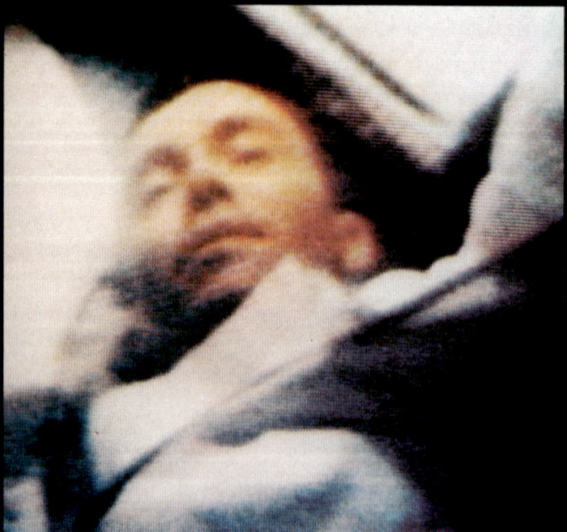

3

4

day later, Mission Control has only a minimum staff (**5**) during the astronauts' fourth routine telecast; Haise is on the screen, explaining a function of the LM. Apollo 13 is now 205,000 miles from earth and about to pass the "point of no return" where the moon's gravity becomes stronger than earth's. Only a few minutes after the telecast, Swigert suddenly shatters the calm with an electrifying message: "Hey, we've got a problem here." Capcom: "This is Houston; say again, please." Swigert: "Houston, we've had a problem. We've had a Main B bus undervolt." The time is 10:10 PM EST. Haise chimes in: "And we had a pretty large bang associated with the caution and warning here." All three astronauts felt the spacecraft shudder. Main Bus B is one of two electrical panels feeding power to the entire spacecraft. The buses draw power from three electricity-generating fuel cells in the SM, which in turn are fed by two liquid oxygen and two liquid hydrogen tanks, also in the SM. These are vital components; the fuel cells also provide water for drinking and cooling. At the time of the emergency they are connected this way (**6**-hydrogen tanks not shown). Less than a minute later Lovell, his voice strained, reports that Main Bus B voltage is now zero, and Main Bus A voltage is dropping. Capcom helplessly requests the astronauts to check connections. Swigert comes back with the chilling news that fuel cells 1 and 3 are now showing no flow. Its power dimming, *Odyssey* is dying in pitiless space. It is Monday, April the 13th.

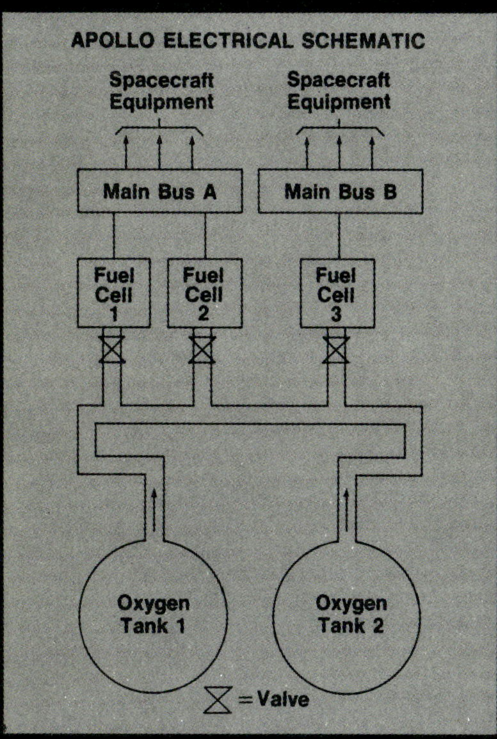

APOLLO 13
In-Flight

An hour after the accident the flight controllers and the astronauts know *what* has happened, but they don't know *why*. Oxygen tank No. 2 has ruptured violently, venting its 300 pounds of oxygen into space and damaging the plumbing so that the other oxygen tank is leaking. Two fuel cells are dead, the third failing; there is no way to fix the damage. The only reserve power in the CM are three re-entry batteries. The moon landing is called off, for the LM is urgently needed now; it will be the astronauts' lifeboat. The crew is told to "power down" the CM—shut off all its switches—and move into the LM, which has its own independent power and oxygen supplies. But a vital question arises. Since Apollo 13 must now go around the moon to get back to earth, can the supplies aboard *Aquarius* be stretched to last three men four days? Houston starts figuring. By 1 AM EST on April 14 the changeover is completed, including the all-important transfer of Apollo 13's navigational co-ordinates from the CM guidance system to the LM's.

1

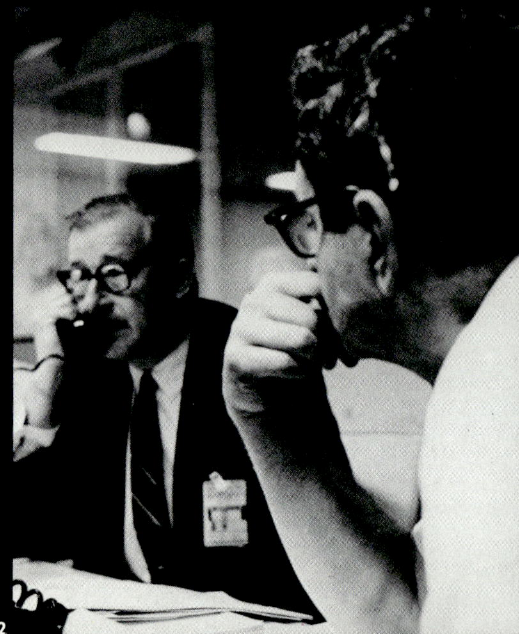

2

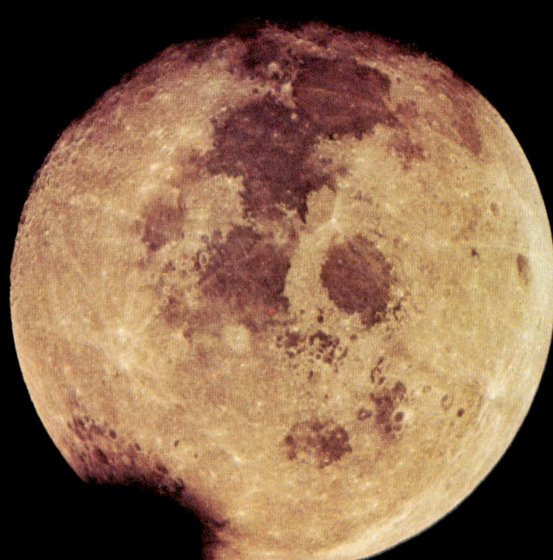

5

6

Now it is a race to keep the astronauts alive. After running alternative courses through computers, Mission Control chooses a small burn now to get Apollo 13 back on a free-return trajectory (otherwise it will miss the earth altogether), followed by a longer burn after the spacecraft has rounded the moon. These burns must be accomplished by the LM's descent stage engine, for not only is the Service Module's SPS engine probably damaged, but it can only be controlled from the shut-down CM. The spindly, spidery *Aquarius*, meant only to shuttle two astronauts to the moon from lunar orbit, must now play locomotive and haul the other two modules back to earth. At 3:43 AM EST the 30-second course correction burn is accomplished, which will bring the astronauts down in the Indian Ocean even if the second burn should fail. During the long night, vigil is maintained at Mission Control (**1**), and NASA Administrator Dr. Thomas Paine (**2**) assures President Nixon by telephone that everything possible is being done. (**3**) Tom Stafford, center, discusses Apollo 13's problems with two of the nex astronauts in space: Apollo 14 Commande Alan Shepard (left) and his LM pilot Ed Mitchell. The man who wasn't there (**4**) Tom Mattingly ponders the fate that sen another man into this mortal danger in his place. The Apollo 13 crew photographs thei onetime destination (**5**) and dark-floored crater Tsiolkovsky (**6**) as they sweep around Mission Control is now a beehive (**7**), as astronauts and controllers jam the consoles

4

On the ground there is intense preparation for alternative maneuvers Apollo 13 may have to perform. Astronauts at Houston and Cape Kennedy run simulators around the clock, for the configuration of the LM pushing the other two modules was tried only once before, a practice run during Apollo 9. A linked LM and CM without the SM has never even been simulated before. The LM's guidance computer is a heavy user of precious power and cooling water, but the astronauts report that the accident has left a cloud of debris around the spacecraft, and they cannot get an accurate star fix. If the computer is shut down, Apollo 13 will lose its bearings; the Flight Director keeps it going. Two hours after the closest approach to the moon (pericynthion plus two hours, or PC + 2), the LM's DPS engine is fired up again. Lovell must control its thrust manually, for the on-board computer has not been programmed for this situation. He warms it up gradually, then applies full power for almost four minutes. The burn is almost

1

2
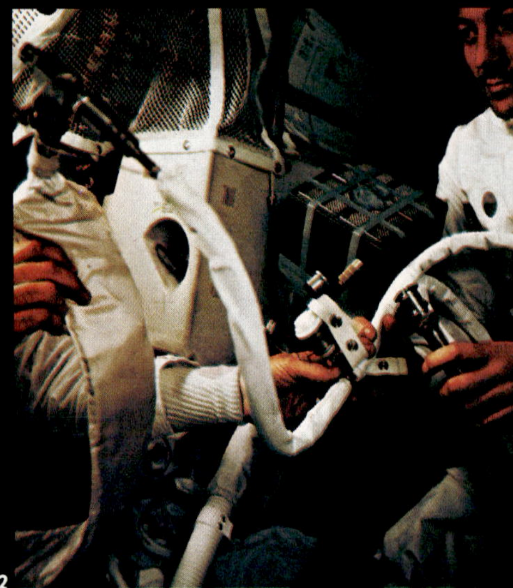

3

perfect—it will require a correction later—and Apollo 13 is now aimed back at the prime recovery area, the southwest Pacific. Unfortunately, a storm is predicted for the area. Mission Control has come up with a way to get a celestial fix using the sun and moon, so the guidance computer is turned off; the astronauts get some sleep. Response to Apollo 13's trouble is heartwarming. Almost every nation with a navy volunteers help, including the Soviets, who offer four ships. But at Mission Control two biological problems need solving. The LM's air-purifying lithium hydroxide canisters will not last, so the astronauts are advised how to jury-rig the CM's canisters (**1**). A urine disposal system (**2**) is also worked out. All other consumables—water, food, oxygen, power—will now last. As the earth looms closer (**3**), the time comes to jettison the Service Module, which has been carried along to protect the CM's heat shield from the intense cold of space. This is done at 8:15 AM EST on April 17, and the astronauts are aghast when they can see the damage for themselves (**4,5**). Lovell: "Right by the high-gain antenna the whole panel is blown out . . . it's really a mess." Now the crew transfers back to *Odyssey*, its batteries recharged from the LM, and bring it back to life. At 11:43 AM EST the astronauts say goodbye to the faithful LM. The hatches are closed, the docking probe uncoupled, and compressed air between the spacecraft gently pushes the LM away (**6**). Capcom sums up everyone's feelings: "Farewell, *Aquarius*, and we thank you."

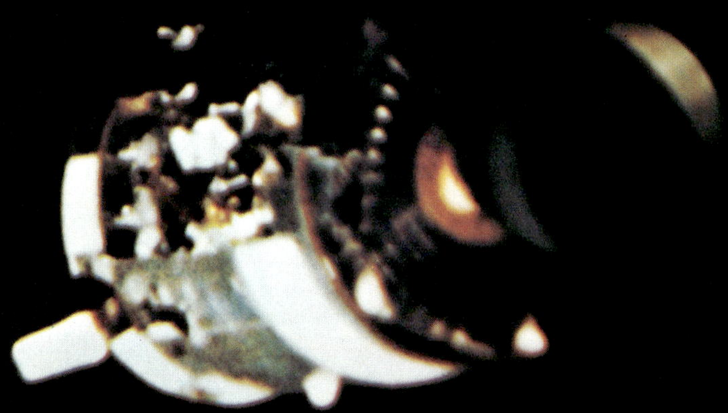

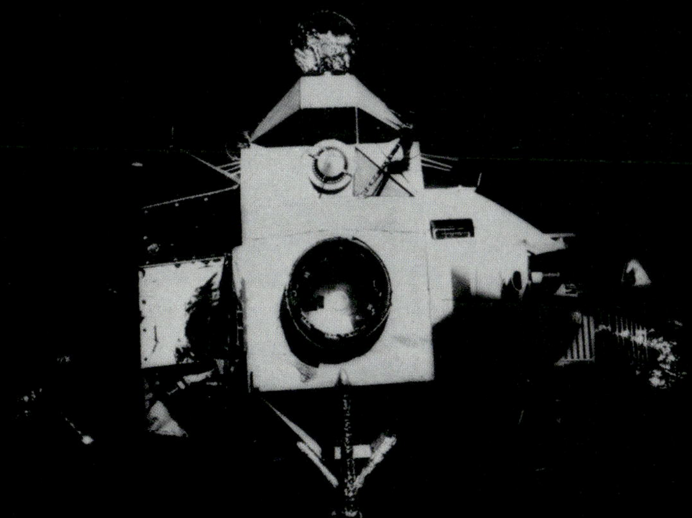

6

APOLLO 13
Splashdown

The new re-entry flight plan and checklist has been laboriously worked on by Mission Control since the PC + 2 burn several days before. Normally, flight plans require months to prepare. But Houston has had round-the-clock help from Cape Kennedy, Grumman in New York, North American Rockwell in California and MIT in Massachusetts. The checklist is "voiced up" to Swigert by Mattingly who spent four hours in a simulator working on it personally; it requires two hours for the former CM pilot to tell his replacement exactly what has to be done. Another problem which concerns the controllers as they pore over meterological maps (**1**) is tropical storm "Helen," which perversely appears headed directly towards the splashdown area. Shift 4 Flight Director Glynn Lunney is seated nearest camera, and Deputy Director of the MSC Dr. Christopher Kraft is standing (in dark suit) at right. Nagging at the Retrofire Officer has been a continual and inexplicable shallowing of Apollo 13's angle of approach ever since

1

2

5

3

7

4

the accident. This situation is now critical at re-entry time, since it affects the Command Module's orientation attitude as it enters the atmosphere. The angle is measured in hundredths of a degree; it is now 6.2 degrees, and the Retro Officer gambles it will not shallow any more. "Lift vector up," he orders, and Capcom informs the astronauts, who fire *Odyssey's* thrusters to properly position it. Of deadly but helpless concern to everyone is the condition of the heat shield. Has it been damaged by the oxygen tank which was below it? There is no way of knowing, and nothing can be done about it now. On board the carrier USS *Iwo Jima* (**2**), preparations begin for recovery amid fog and rain; "Helen" is nearby. But as *Odyssey* plunges back into the atmosphere it left six days before, the storm veers away. Through a break in the clouds (**3**), the spacecraft is finally spotted, dangling safely from its three big parachutes. An immense sigh of relief is breathed everywhere, but especially at Mission Control (**4**), where these two controllers stick to their posts to the end; practically everyone else is at the rear of the room, preparing for the celebration. At 1:07:44 PM EST, *Odyssey* hits the Pacific only four miles from the *Iwo Jima.* A few minutes later, an elated frogman (**5**) gives the traditional "A-OK" sign. Cigars, smiles, and applause break out in crowded Mission Control (**6**); NASA Administrator Dr. Thomas O. Paine (center) is perhaps happiest of all. Mission Control was never so jammed or jubilant (**7**) as this moment.

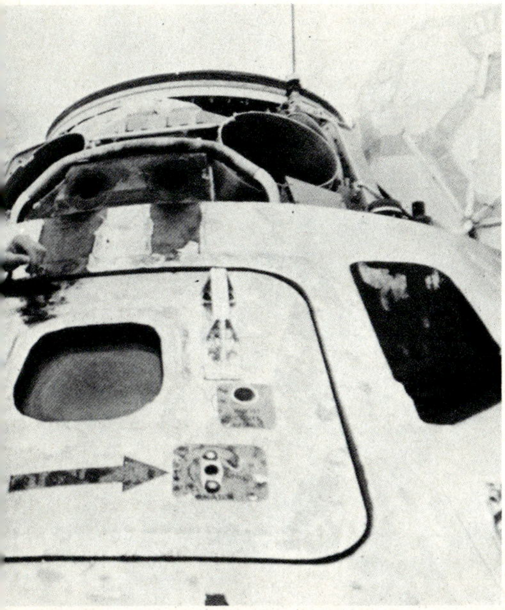

6

APOLLO 13
Splashdown

Very tired but very happy, the astronauts get the red carpet treatment (**1**) aboard the *Iwo Jima*. Lovell, a Navy Captain himself, is about to salute a superior officer, but Rear Admiral Donald C. Davis just extends his hands. "Welcome home, Jim. Well done." Since the astronauts did not land on the moon, they don't have to go into the isolation van. While below for rest and medical examination, they receive the traditional telephone call from the President (**2**). Swigert apologized for his income tax form being late—in the last-minute pressure of the crew switch, he had forgotten to file it by April 15th. But it is laughingly pointed out to him that since he was out of the country—very far out—he is eligible for a 60-day extension. *Odyssey*, which truly had a journey worthy of its name, is plucked from the sea (**3**), after which the ship sails for the island of Pago Pago in Samoa. Back in Houston, Mrs. Marilyn Lovell holds a little press conference (**4**) outside her home after the successful recovery of her husband. With her are three

1

2

3

of the Lovell children: Susan (left) and Barbara, holding Jeffrey. Aboard the *Iwo Jima* , Jim Lovell reads all about it (**5**) in a Honolulu newspaper flown in from Hawaii. He looks surprised to learn he has landed safely. The medical examination shows the Apollo 13 crew to be in much worse condition than any previous returning astronauts. Their reflexes are slow, they are unsteady on their feet, and they are exhausted. In the 3½ days since the accident, they averaged only 3 hours of sleep a day. The sleep they did manage was poor, because the spacecrafts were so cold. It was bad enough in the cramped LM, but if an astronaut "went upstairs" to stretch out on one of the CM's couches, he almost froze; the temperature eventually dropped to 35 degrees in *Odyssey.* Haise has a kidney infection, and Swigert has near-frozen feet. All three astronauts are dehydrated, because they were limited to only six ounces of water a day, a fifth of their normal consumption; the inoperative fuel cells normally provided water, and a spare water tank in the CM had frozen solid. Their levels of body fluids and andrenal steroids were down significantly. If the flight had lasted much longer, the astronauts would have drifted off into sleep. From Pago Pago the crew is flown to Hawaii, greeted there by President Nixon, NASA officials, and family members. They are each awarded the Presidential Medal of Freedom (**6**), and of them Mr. Nixon says, "They did not reach the moon, but they reached the hearts of millions of people."

4

5

6

APOLLO 13
Post-flight

President Nixon (**1**) also awarded the Presidential Medal of Freedom to the Director of Flight Operations, Sigurd A. Sjoberg, and to the four Flight Directors who alternated positions around the clock: Glynn S. Lunney, Eugene F. Kranz, Gerald Griffin, and Milton L. Windler. Upon his arrival back in Houston, Lovell greets Mattingly warmly (**2**). "Do you have the measles yet, Tom?" "Not yet, Jim; you scared them right out of me." This is oxygen tank No. 2 of Apollo 13 (**3**), a photo taken before launch. The two oxygen tanks were located in the middle of Bay 4 of the SM, between the three fuel cells and two hydrogen tanks. Inside each tank are heaters, "fans" for stirring, and sensors, all fed by wires through the dome on top. The accident was eventually traced to mishandling of the tank during the Countdown Demonstration Test in March. Technicians had trouble releasing the oxygen after the CDDT, and tried to force it out by turning on the fans and heaters. It worked, but the heaters were left on for eight hours. The thermostat-

1

2

ic safety switch was made to operate on the CSM's supply power of 28 volts, but the tanks were tested at 65 volts. As the oxygen level dropped, this switch was no longer cooled as intended by immersion. The overload voltage plus lack of cooling fused the switch shut, and tank temperature soared to perhaps a thousand degrees, enough to melt the wiring insulation inside. Just before the accident, Swigert was routinely requested by Mission Control to turn on the fans. In oxygen tank no. 2 an arc jumped between two bare wires. The oxygen heated rapidly, and in less than a minute the tank burst apart, wrecking the other oxygen tank and blowing off the door on Bay 4. Another view (**4**) shows the hydrogen tanks with the two oxygen tanks above them. Future Apollo spacecraft will be modified by stainless steel sheathing for the tank wires, removal of the fans, and installation of an additional oxygen tank, battery, and 5-gallon water supply. The aftermath was not without corporate humor; Grumman presented North American Rockwell with a 300,000 mile towing charge. Apollo 13's S-IVB lunar seismology crash went well, and the curious shallowing of the return trajectory was found due to venting of the LM's cooling water. Facts of the mission were presented to the Senate Space Committee (**5**) on April 24 by (left to right) Dr. Rocco A. Petrone, Dr. Thomas Paine, Jim Lovell, and Jack Swigert. Twice to the moon, almost to his death in space, Jim Lovell was quietly eloquent: "We do not realize what we have on earth until we leave it."

4

It is probably quite natural for the focus of attention in our space program to be concentrated on the gigantic Saturn rocket and the men inside it. True, a more impressive focal point would be hard to find. But the launch vehicle and astronauts are like the tip of an iceberg. They represent the small, visible part of a large and complex system. In the background is a lesser known but equally impressive and crucial aspect of the space program: The Manned Space Flight Network. It is the lifeline (**1**) that is the astronauts' electronic umbilical to earth. Even science fiction writers like Jules Verne, who predicted man's landing on the moon with great insight and detail, did not visualize almost half the people on earth watching the event as it was happening. That minor miracle was courtesy of the Manned Space Flight Network. The Apollo 13 astronauts also owe a great deal to the NASA Communications Network (**2**), of which the Manned Space Flight Network is only one part. Although the network is out of touch

APOLLO COMMUNICATIONS LINK

"Any chain is only as strong as . . . ."

1

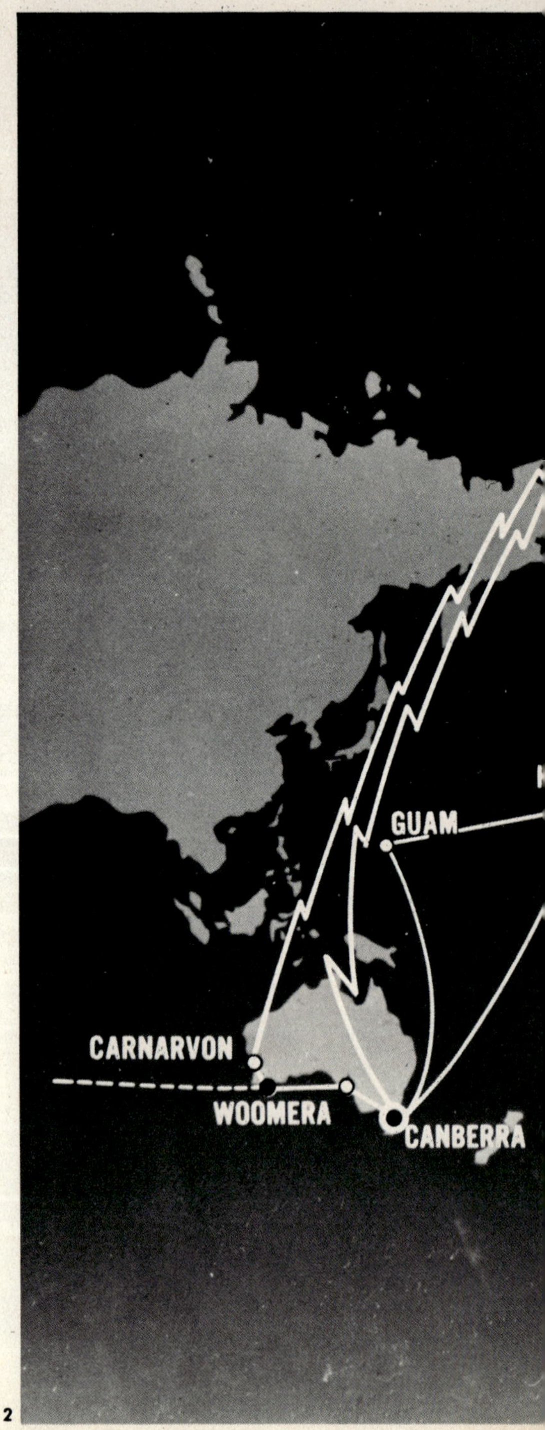

2

with the astronauts only when the spacecraft is behind the moon, the Communications Link has proved to be incalculable. Without it, Apollo 13 surely would have been lost. For, in the vastness of space, the pleas of that tiny object crackled around our world: "Hey, we've got a problem here." Through that communications network, a reassuring word was instantaneous: "This is Houston, say again, please." Thus began one of the most epic feats of communication in the history of man. *Spacecraft:* "Okay, Houston. . .I tried to reset, and fuel cells 1 and 3 are both showing zip on the flows." *Capcom:* "We copy." *Spacecraft:* "Houston, are you still reading 13?" *Capcom:* "That's affirmative. We're still trying to come up with some good ideas here for you." The world held its breath; and step by step through the long night, a daring solution was relayed to the astronauts from the ground. The Manned Space Flight Network (MSFN)—having ten years, 4000 men and $600 million with which to build—shone as a bright star in that dark night of Apollo 13, and far exceeded its value. The MSFN uses 17 ground stations, including prime sites in Australia and Spain; and the 210-foot antenna system at Goldstone, California. Six to eight instrumented aircraft and four ships complete the network. The MSFN is not only required to handle the total audio and video load, but with its sophisticated computer can unscramble and assimilate 51,000 bits of information for each second of spacecraft flight, equivalent to 50 typewritten pages a second.

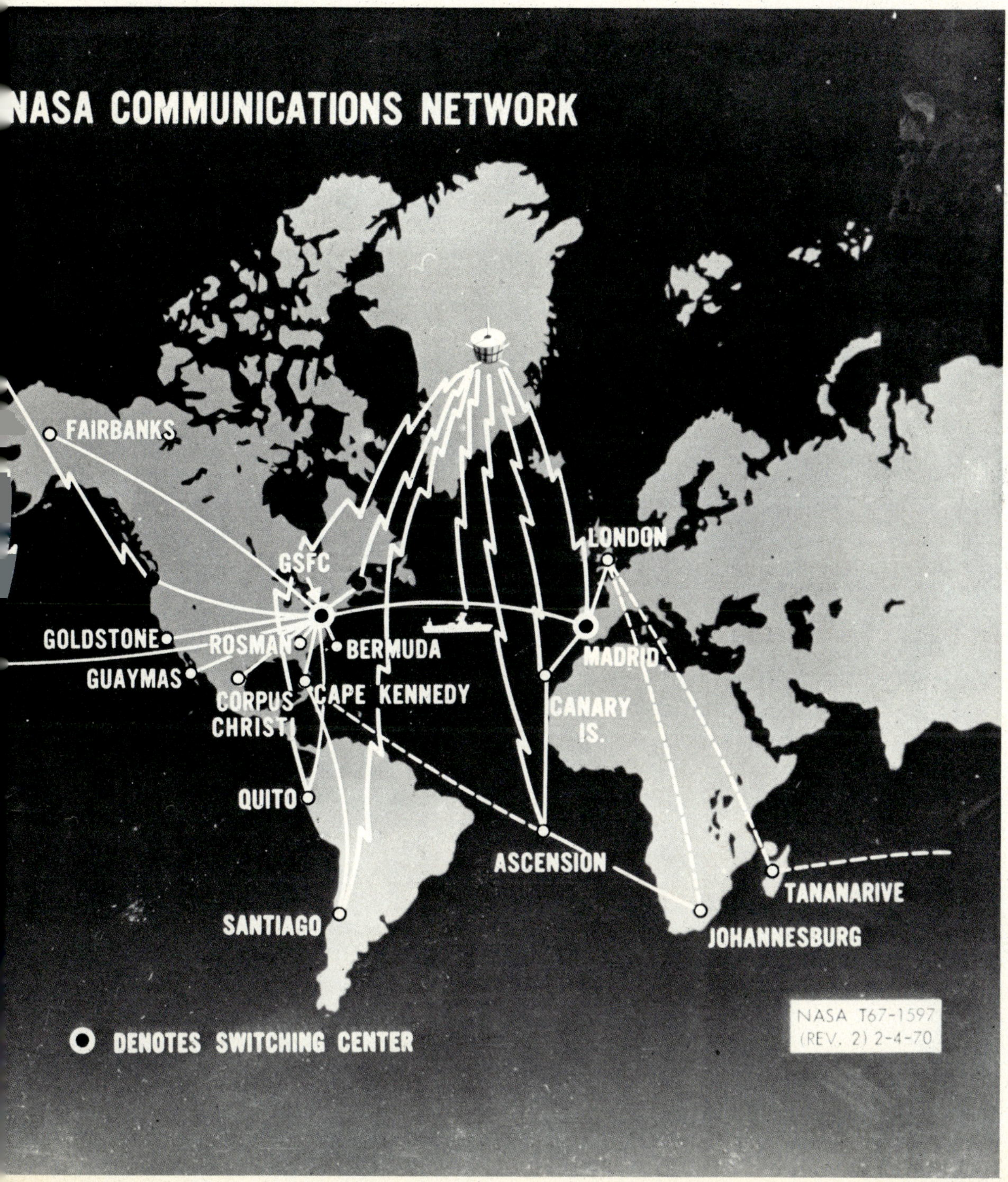

COMMUNICATIONS

The scrambled or "bundled" messages from orbiting spacecraft are transmitted in one carrier wave. After they are "unbundled" and sorted out by computers, they are immediately transmitted to Houston and Goddard. Conversely, commands from Houston are checked, processed, and transmitted to the appropriate Manned Space Flight Network station for relay to the spacecraft. Communications for the Apollo missions—from a spacecraft to ground and vice versa—are transmitted at a speed up to 80 times greater than those for the Mercury missions. The Goddard Space Flight Center at Greenbelt, Maryland (**1**), was established by NASA on May 1, 1959. It was the first major laboratory devoted entirely to the investigation and exploration of space. Located about ten miles northeast of Washington, D.C., Goddard is the control center for the entire NASA Communications Network (NASCOM), in addition to its primary responsibilites for experimentation in tracking, data acquistion, and interpretation. NASCOM

1

2

3

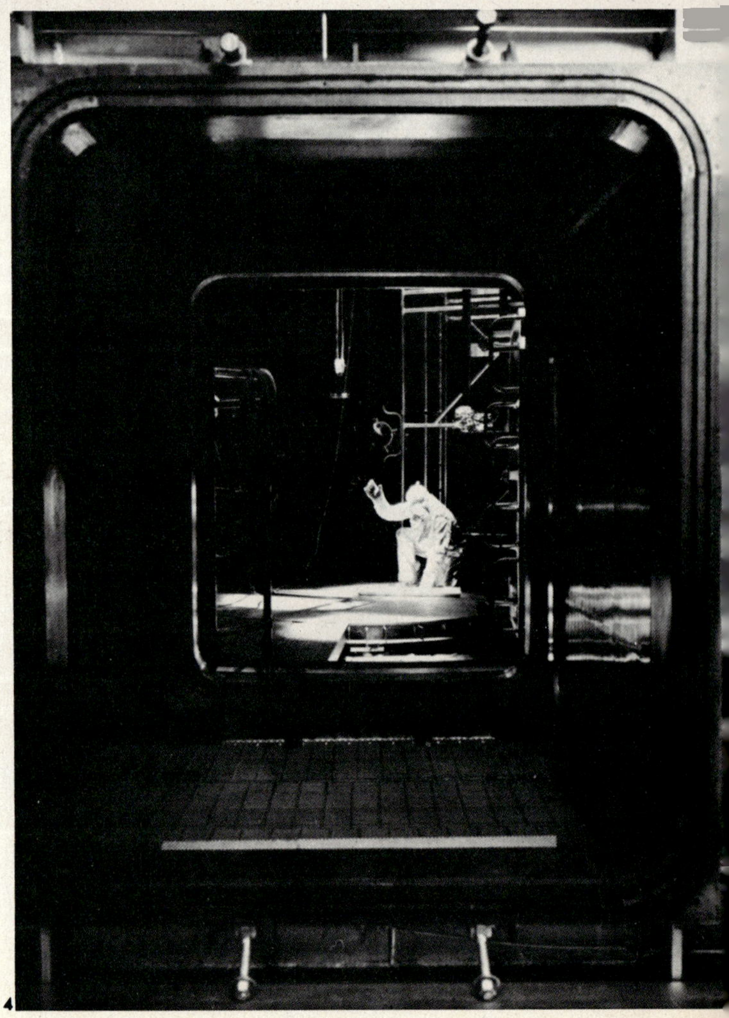

4

is the most expensive and sophisticated communications network in existence, consisting of nearly three million miles of diversely routed communications channels. The system uses satellites, submarine cables, land lines, microwave stations, and high frequency radio facilities to provide the necessary access links. Here (**3**), a telemetry antenna system test is being conducted on a proposed Supersonic Planetary Entry Decelerator. This test is taking place in the anechoic (low sound) chamber at Langley Research Center, Hampton, Virginia. Looking more like an antenna, the decelerator actually works like a large umbrella. An artificial sun glares downwards, as a technician works inside the Space Environment Simulator (**4**). He is wearing a special reflective work suit to protect against ultraviolet rays. Another Goddard activity is the Network Training and Test Facility, which duplicates a complete tracking station. The facility trains both personnel and systems (**5**), and has its own 30-foot tracking antenna (**2**). In addition, there are sea-going tracking facilities, such as the USNS *Vanguard* (**6**). The *Vanguard's* primary role in the Apollo program is tracking and relaying data during earth orbit insertion. The Goddard Space Flight Center is joined by regional communication switching centers in London, England; Madrid, Spain; Canberra, Australia; Honolulu, Hawaii; and Guam island. During Apollo missions, the focal point for all flight control activities (**7**) is Houston, with the NASCOM relaying all data there.

6

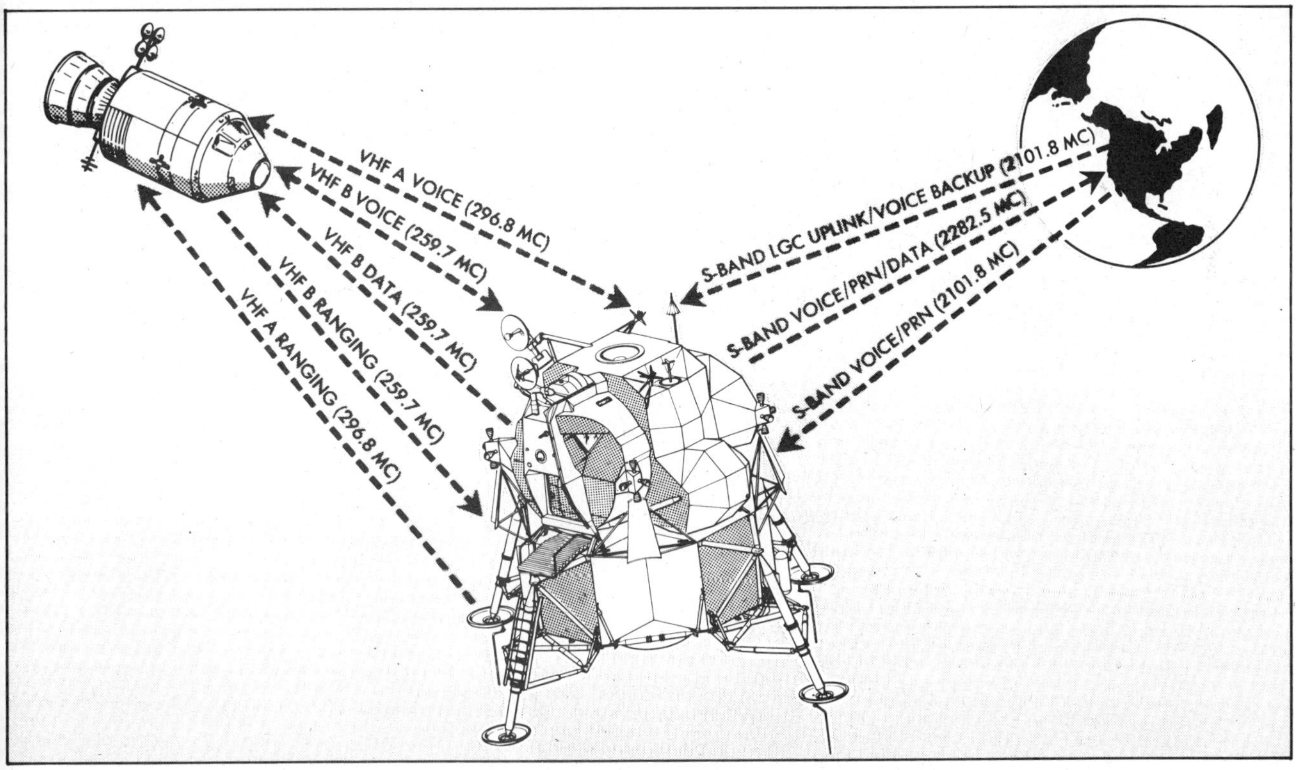

COMMUNICATIONS

The end result of NASCOM is reflected in abilities of the farthest link in the lunar mission program. That, of course, would be the Lunar Module, and ultimately the exploring astronauts. But, until the Lunar Roving Vehicle is introduced with Apollo 15, all communications must leave the moon through the LM's network of equipment, the S-Band steerable antenna at the top of the drawing (**1**). A view of the Madrid tracking station shows individuals at work transmitting and receiving information at lunar distances (**2**). The stations of this type are located at 120-degree intervals around the earth. They are arranged this way so that at least one of the 85-foot antennas will always be in "view" of the spacecraft during a lunar mission, despite the rotation of the earth. Goldstone, however, has both 210-foot and an 85-foot-diameter antenna. Thus, with this vast and expensive communications network, the world could hear the instant excitement from the explorers of Apollo 12, catching a precious moment of discovery. At one point,

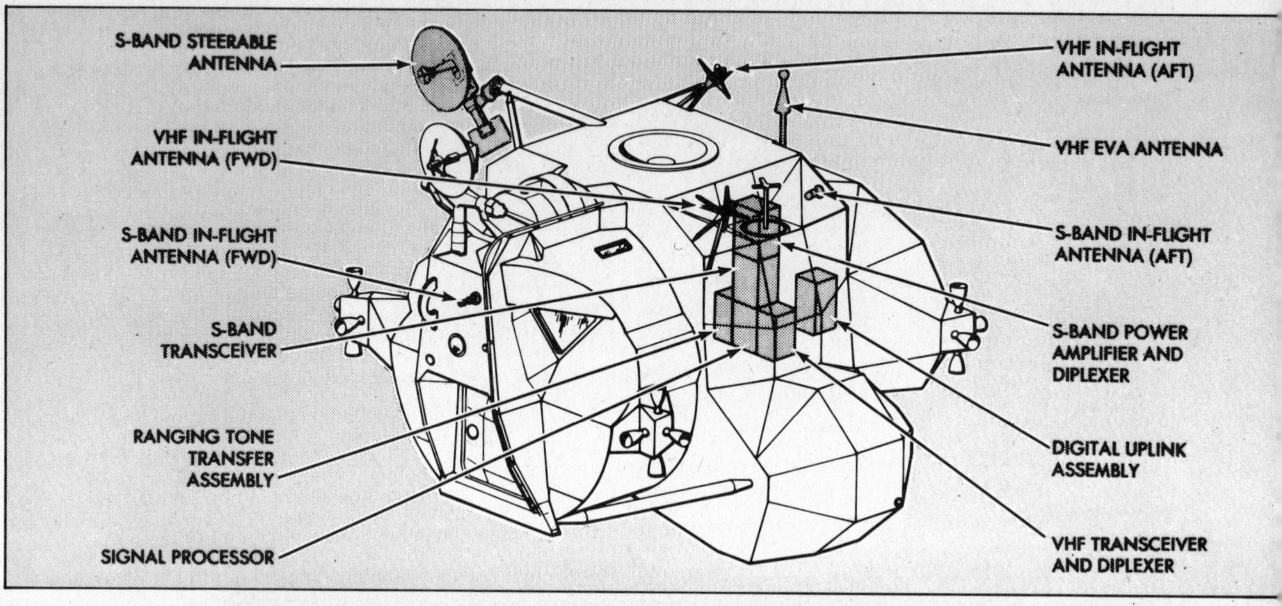

1

4

the astronauts had just walked by a mysterious mound of lunar soil. After warning Houston not to take the following comment the wrong way, Alan Bean said: "It looks like a small volcano, only it's just about 4 feet high and about 5 feet across the top. It slopes down to a base with a diameter of 15 to 20 feet." But with immediate communication capability, geologists on earth were able to speculate that the mounds were piles of coarse materials ejected from a crater. The astronauts on the scene were thus spared the need for further delay, by the reassuring voice link with earth. Another key to vital air-to-ground voice relay links between an Apollo spacecraft and the Mission Control Center is the fleet of eight specially modified KC-135 4-engined jet aircraft, which carry Apollo communications equipment and operator specialists (**3**). These eight aircraft, operating on a world-wide basis, use airfields in many countries for staging, refueling, and recovery. Not only are these long-range aircraft loaded with sophisticated cameras and tracking gear, but each is equipped with a nose radome (**7**), housing a 7-foot parabolic S-Band antenna for voice relay and receiving. The Goldstone 85-foot antenna (**4**) is in position for use; while at sea in a distant corner of the earth one of the four Apollo tracking support ships (**5**) displays its considerable variety of dish, helix and whip antennas. An important part of NASA's world-wide communications network is this tracking and data acquisition ground station (**6**) close to Johannesburg, South Africa.

3

7

APOLLO 14

Apollo 14 is a child of destiny, NASA's response to a near disaster. The flight was originally scheduled for launch in October 1970, its destination the crater Littrow near the Taurus Mountains. But the Fra Mauro region for which the ill-fated Apollo 13 flight was intended is of such great interest to selenologists that Apollo 14's target is shifted to it. The significant modifications ordered to spacecraft as a result of Apollo 13 twice cause a postponement of the Apollo 14 launch, finally scheduled for January 31, 1971. This mission will inaugurate several flight plan variations. Lunar orbit for the CSM will be much more eccentric in order to launch the LM at a lower altitude, and the LM's ascent stage will attempt rendezvous by intercepting the CSM directly from lunar launch rather than by matching orbits. A wheeled tool carrier will be used, and two ALSEP experiments are new. An "uprating" of the PLSS backpack will allow a much greater EVA time. The crew named their CM *Kitty Hawk* after the North Carolina site of the Wright brothers' famous flight in 1903 because, as they put it, "That's where it all began." Their LM *Antares* is named for the bright red star in the constellation Scorpio, on which the spacecraft will orient itself during descent to the moon.

ALAN B. SHEPARD, JR.

SPACECRAFT COMMANDER/ One of the original and best-known astronauts, Shepard's suborbital flight of May 1961 made him the first American in space. Grounded with a disorder of the inner ear, he became Chief of the Astronaut Office in 1963. Following surgery, he was restored to flight status in 1969, and yielded his office to Tom Stafford when named Commander of Apollo 14. His quick wit, impish humor, and still-boyish appearance belie his 47 years. This flight is his dream come true—it's been a long time.

STUART A. ROOSA

COMMAND MODULE PILOT/ Born 37 years ago in Durango, Colo., carrot-topped Stu Roosa grew up in Oklahoma and moved around the West a good deal before graduating from the University of Colorado with a BS degree in Aeronautical Engineering. Prior to becoming an astronaut in 1966, he was a fighter pilot at several bases and then a test pilot at Edwards AFB, Calif. Now a Lt. Col. in the USAF, he has acquired 4900 flying hours, almost all in jets. He and his wife have four children, and he enjoys outdoor sports.

EDGAR D. MITCHELL

LUNAR MODULE PILOT/ A Navy Captain, Ed Mitchell was born in Hereford, Texas, on September 17, 1930, but calls the New Mexico town of Artesia his home. He entered the Navy in 1952, and holds a BS degree in Aeronautical Engineering from the Naval Postgraduate School, as well as a Doctorate of Science in Aeronautics/Astronautics from MIT. Before becoming an astronaut, he was a test pilot and carrier flight officer, 4000 hours flying time. The backup LM pilot for Apollo 10, he and wife Louise have two children.

APOLLO 14
Shepard
Roosa
Mitchell

APOLLO 14
Preparation

In order to ensure that all the members of the prime and backup crews are completely versed in the intricacies of living in and operating the spacecraft, much of their training takes place inside a Command Module Simulator. This is a replica of the actual Command Module, and is computer-controlled to respond to inputs as the real one will in flight. By the time the crew is ready to launch, little has been left to chance or guesswork. At the Kennedy Space Center, Florida, (**1**), the crew participates in Command Module simulation. From left to right: Edgar D. Mitchell, Lunar Module pilot; Stuart A. Roosa, Command Module pilot; and Alan B. Shepard, Jr., Commander. These seemingly endless simulations are an integral part of an astronaut's life, and are designed to make virtually every facet of the mission as close to a conditioned reflex as possible in light of the job's complexity. A fish-eye view (**2**) shows Shepard (foreground) and Mitchell in the Apollo Lunar Module Mission Simulator (LMMS) at the Kennedy Space Center

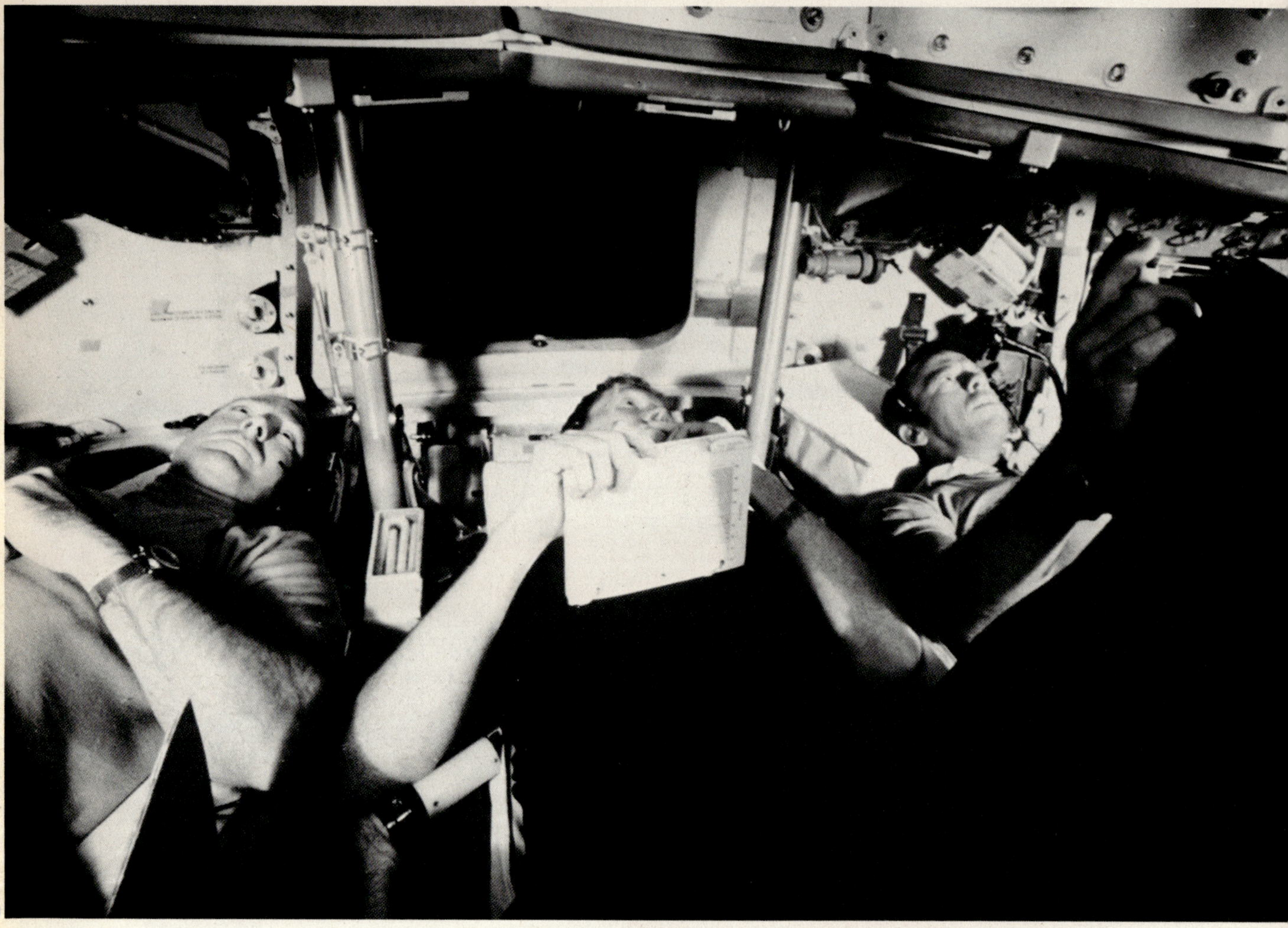

1

3

during pre-flight training, familiarizing themselves with the machinery and equipment they will be using during the actual mission. Here (**3**) Mitchell demonstrates something new on Apollo missions—the wheeled tool carrier, officially designated by NASA as the Modular Equipment Transporter and quickly dubbed "rickshaw" by the astronauts. Prior experience of Apollo 11 and 12 missions has revealed the fact that, despite the moon's 1/6 gravity, working on the lunar surface is very strenuous because the pressurized EVA suits tend to have a balloon-like stiffness and the astronauts must expend extra effort to flex the bellows-like joints at the knees and elbows when deploying the ALSEP, picking up lunar soil samples or merely moving across the boulder-strewn lunar terrain. Thus, to conserve energy and add valuable exploration time, MET was born. An artist's concept (**4**) shows how the MET will be used on the lunar surface, while (**5**) shows a prototype. The "rickshaw" was used by the Apollo 14 astronauts during their geological and lunar surface simulation training in the Pinacote volcanic area of northwestern Sonora, Mexico; they will be the first ones to use the MET. Weighing 20 lbs, it will be a portable workbench with places for the lunar handtools (including 6 core tubes, a trenching tool, two geology scoops and a portable magnetometer to measure the moon's magnetic field) and their carrier, three cameras, two sample container bags, a environmental sample container, spare film magazines and a soil surface penetrometer.

4

5

APOLLO 14 Preparation

As the launch date for Apollo 14 draws nearer, it is more inportant than ever that the astronauts be prepared to function competently under lunar conditions. Since it is obviously impossible for the team to train on the moon, the training conditions must simulate as closely as possible those existing on the lunar surface. Although the moon has no atmosphere, it is not necessary for the astronauts to train in a vacuum; provided the space suits are thoroughly tested for pressure integrity, merely wearing them on earth will duplicate wearing them on the moon. Finding similar terrain is no real problem but approximating the temperatures is: it will be well over 200 degrees, hotter by far than any place on earth. Most difficult to reproduce is the gravity difference. Because the moon's gravity is 1/6 that of earth, it is important for the astronauts to train in an environment that artificially duplicates the lessened gravity, wearing the space gear and clothing they will wear on the moon. (**1**) On December 14, 1970, Shepard begins a

1

2

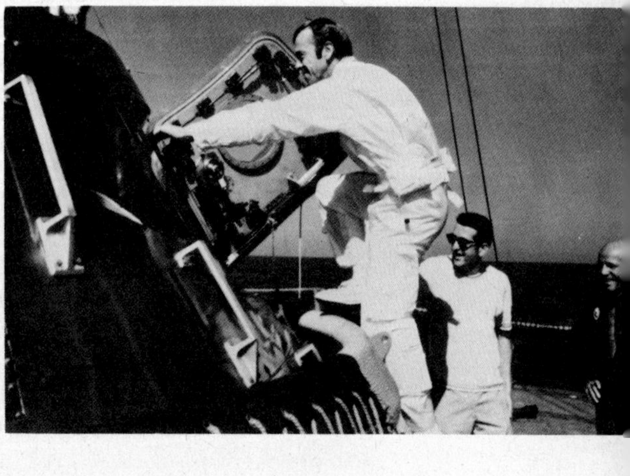
3

test flight in the Lunar Landing Training Vehicle (LLTV) at Ellington Air Force Base, Texas, which is quite close to the Manned Spacecraft Center. The ungainly-looking "flying bedstead" maneuvers like a Lunar Module in a lunar environment, and Shepard and Mitchell spend many hours in it, as did their predecessors on previous missions. Shepard will land the real Lunar Module, and Mitchell will assist. At Patrick Air Force Base, Melbourne, Florida, (**2**), backup LM pilot Joe H. Engle takes part in a 1/6 gravity test. The particular test shown here was designed to obtain an ingress-egress evaluation of the LM under 1/6 gravity conditions. The test is conducted aboard a KC-135 aircraft flying a parabolic flight path that negates 5/6 of the earth's gravity. Even though Engle is not a member of the Apollo 14 prime crew, all the astronauts involved in the mission undergo the same tests. Being able to get in and out of the space craft is of a vital function, and here, (**3**), Alan Shepard practices getting into (ingress) a Command Module trainer aboard the NASA Motor Vehicle Retriever, during preparations for water egress training in the Gulf of Mexico. Knowing how to pull a 20 lb. cart behind oneself is a fairly simple operation on earth, but even this motion has to be rehearsed. Aboard a KC-135 again (**4**), Shepard pulls the MET rickshaw while Mitchell brings up the rear. The two lunar surface astronauts also rehearse deploying tools from the MET under 1/6-gravity conditions to know exactly how much energy each move will take.

APOLLO 14 Pre-Launch

While the astronauts prepare for working on the moon, the huge and complex machine that will take them there is assembled. On January 21, 1970, a year before launch, the S-II second stage (**1**) of the Apollo 14 space vehicle arrives at Kennedy Space Center and is off-loaded at the Turning Basin. The stage is then trucked into the nearby Vehicle Assembly Building (VAB) at Launch Complex 39, where it will be stacked into the Saturn V rocket. This stage was built by North American Rockwell Corporation in California, and is now being brought by sea from NASA's Mississippi Test Facility, where it was test-fired. The S-IC first stage (**2**) is shown being erected in High Bay #3 of the VAB. Rollout of the Apollo 14 launch vehicle A/S 509 (**3**) began at 6:30 am EST, Nov. 9, 1970, and took about 5 hours for the 3½ mile trip from the VAB to Pad 39A. Apollo 14's CSM is spacecraft #110; *Antares* is Lunar Module 8. On January 19, 1971, the Countdown Demonstration Test proceeds smoothly, much to everyone's satisfaction

1

4

5

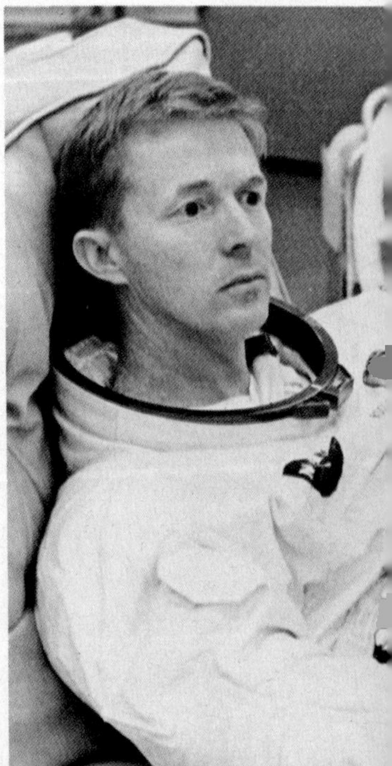

6

after the near-disastrous Apollo 13 mission. On launch day, Jan. 31, the crew enjoys the traditional sendoff meal (**4**), this time called a brunch rather than breakfast since it took place in mid-morning. Around the dining table in the Spaceport Crew Quarters are (facing the camera, left to right): Eugene Cernan, backup Commander; Mitchell; Tom Stafford, Chief of the Astronaut Office, a position formerly held by Shepard; and Roosa. Seated on the left side of the table, from rear to front, are: Ron Evans, backup LM pilot; Stafford's boss Donald Slayton, Director of Flight Crew Operations; and Shepard. Deke Slayton, of course, like Shepard was one of the original seven astronauts, and Stafford still maintains his astronaut flight status. LM pilot Ed Mitchell is shown here (**5**) adjusting his watch with a certain apprehension during suiting-up activities, while Stu Roosa (**6**) seems overcome by it all. Not so the fearless leader. Though his lunar express is due to leave in a few hours on a journey of a quarter of a million miles, the waggish Shepard seems to be telling a fish story (**7**) to a skeptical technician, probably proclaiming that the crew is going to find moon trout "this big". Stu Roosa is shown here (**8**) stepping into the transfer van prior to launch, followed by Mitchell; Shepard is already inside. Technicians and personnel in the background are carrying additional gear. The van transports the crew eight miles to their gleaming white rocket sitting on Pad 39A, which will be their home away from home for the next ten days during the trip.

3

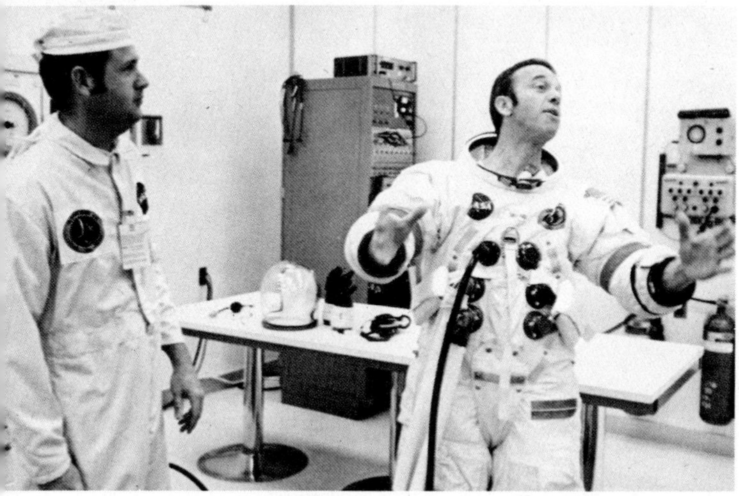

8

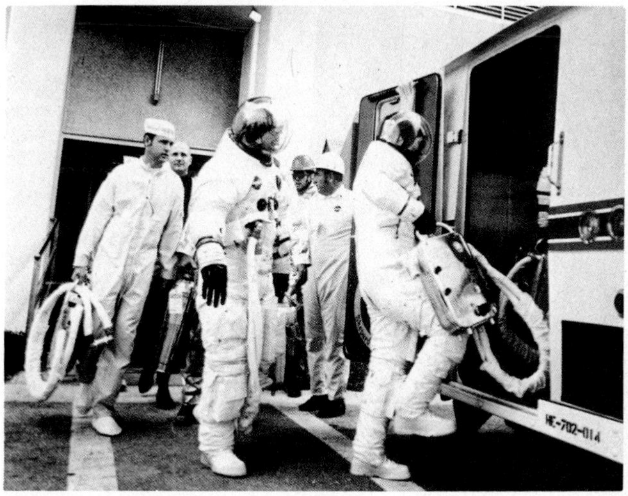

APOLLO 14
Launch

1

2

3

4

Up in Firing Room 2 of Kennedy's Launch Control Center (1), longtime friends and associates Dr. Wernher Von Braun, NASA's Deputy Associate Administrator for Planning, and Dr. Kurt H. Debus, Director of the Kennedy Space Center, discuss the Apollo 14 pre-launch activities, perhaps comparing them to the good old days at Peenemunde. The Spanish and American flags fly together (2) over the viewing site to note the visit of Their Royal Highnesses the Prince and Princess of Spain. Princess Sophia, seated between the flags, is waving; she is flanked by Vice-President Spiro T. Agnew and Neil Armstrong, the first astronaut to land on the moon. Prince Juan Carlos is seated next to Armstrong, and below Agnew can be seen presidential advisor Henry Kissinger. One of the most excited and personally involved spectators (3) at the launching is Alan Shepard's wife Louise (under umbrella). She and her friends were in a car for protection from the rain but stepped out to watch the liftoff. Though delayed by weather, the Apollo 14 countdown and launch proceed flawlessly. At 4:03 PM EST the giant rocket billows smoke and flame; a remote camera atop the 380-foot Launch Umbilical Tower catches the moment of ignition, an instant before the service arms swing away. Climbing through the clouds on its fiery tail, the rocket holds the multitude spellbound (4), the crackling thunder of its voice rolling over them like a Biblical revelation. Apollo 14 reaches parking orbit without incident, and after three tours of the earth the S-IVB engine is restarted at 6:37 PM EST to push the crew on their long trip to the moon. At 10:39 PM EST on Feb. 1, a transfer maneuver is performed for a better lunar approach, as was done on Apollo flights 12 and 13. Also like Apollo 13, the S-IVB stage trails along after separation and is aimed to impact on the moon for a seismological test. After 82 hours, 37 minutes of flight, the Service Module's SPS engine comes to life at 2:00 AM Feb. 4, and burns for 6 minutes, 12 sec. This brakes the spacecraft and inserts it into lunar orbit. Forty-one minutes later the S-IVB crashes into the moon at more than two miles per second, setting off shock waves that continue to vibrate for 3 hours. This confirms the similar test results from the two previous missions; the moon indeed has an internal structure much different from earth, and selenologist are both pleased and excited by this data.

In all previous manned lunar flights, the Command/Service Module approached no closer than 70 miles to the moon's surface, at which point the LM would detach from the CSM to begin its descent. Due to the hilly terrain at the Fra Mauro landing site, a different and closer approach will be used for this mission. Apollo 14's lunar orbit is elliptical, 70 miles above the moon at its lowest point (pericynthion) and about 193 miles at its highest altitude (apocynthion), identical to the procedure used on previous missions. After two such orbits, however, a short burst by the SPS engine at 6:14 AM EST decelerates the craft still more, lowering its orbital path to about 50,000 feet above the moon at its lowest point, and about 70 miles at apocynthion. This orbital change-over maneuver is known as "descent orbit insertion" or DOI. By bringing the CSM to this lower orbit and letting *Antares* proceed from there toward a landing, precious fuel is saved, giving Shepard about 14 more sec-onds of hover time to select the most favor-

1

2

3

4

able touchdown point in the hilly target area. After orbiting for 17½ hours in this manner, which gives ample time for powering up the LM and preparing it for landing, at 11:51 PM EST Shepard and Mitchell undock *Antares* . This view of crescent earth (**1**) is seen as the LM approaches the moon prior to Powered Descent Initiation. PDI takes place at 4:05 in the morning of Feb. 5, and 12 minutes later *Antares* settles safely down into the lunar soil, 108 hours and 54 minutes after leaving earth. Shepard and Mitchell are jubilant but tired. A solar reflection forms a bright ball of light (**2**) above *Antares* as it rests on the lunar surface. To the left of *Antares* is the lower slope of Cone Crater. (**3**) A view of the central station of the ALSEP stands out clearly on the lunar surface. Here (**4**) Shepard poses beside the flag he planted during the first EVA; shadows of the LM, Mitchell, and the S-band antenna can also be seen. (**5**) In the foreground here is one of the ALSEP experiments, the Suprathermal Ion Detector, which detects particles generated by the exhaust of the LM's descent engine and later by the ascent engine when they leave the Moon. The MET and ALSEP components can be seen in the background. The scientists involved in the Apollo 14 mission have said that the instruments taken on this mission were the best ever devised, and expect them to continue transmitting information for at least a year. With two seismometers now operating on the moon, it is easier for scientists to locate the source of lunar quakes.

5

There's a long, long trail a-winding here (**1**), but these tracks are not those of an old wagon train across the Mojave Desert; they are the tracks of new pioneers, across the barren, billions-years-old moon, left by the MET when the astronauts pulled it away from the Lunar Module during their second EVA. *Antares* squats in the background, the inverted umbrella of the S-Band antenna beside it. Near the end of Apollo 14's second EVA some unexpected events take place—unexpected, that is, to Mission Control in Houston and the vast television audience which is watching a live transmission. Shepard chortles, "OK, Houston, watch this," and in full view of everybody (**2**) produces a real golf ball, which he then proceeds to drive several hundred yards with a real six iron! There are actually two balls; Shepard drove both of them, missing only with his first swing. The bulky space suit prevents him from getting a proper two-hand grip, but even with his awkward one-hand grip the ball sails a long way in the light gravity; a

2

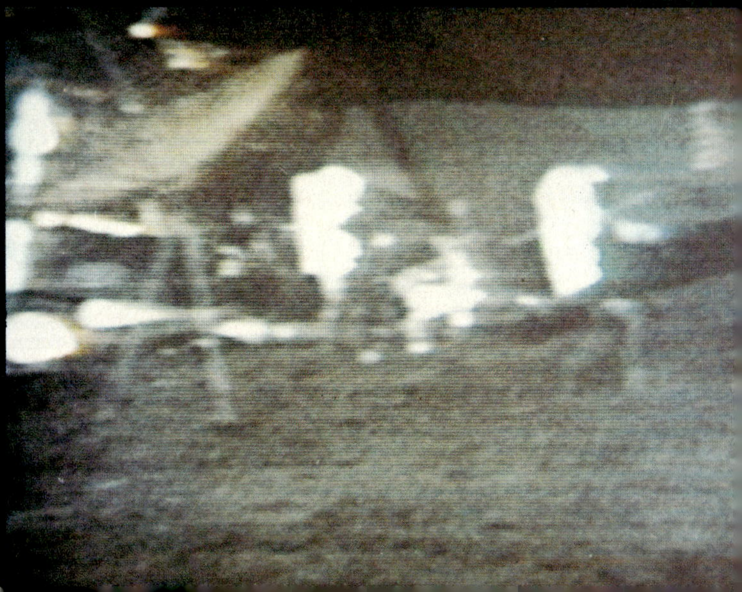

3

300-yard drive on earth would travel a mile on the moon. Shepard's club is an actual six-iron head attached to the end of the handle for the contingency sample return. While the astounded flight controllers are still abuzz over this unprogrammed incident and wondering how Shepard smuggled the golf balls aboard, up steps lunar decathlon contestant Ed Mitchell (**3**) with his act. Using the staff of the Solar Wind Composition Experiment, Mitchell heaves a javelin toss—left-handed, yet. Again the light gravity makes a championship performance despite the restriction of the space suit. While the lunar Olympics are taking place on the surface, Stu Roosa in orbit is taking pictures. One of them (**4**) is a view across Fra Mauro and Parry craters, showing a formation known as the Rima Parry. Note how two of the features which resemble sunken riverbeds apparently emanate from one crater, while another in the upper left crosses over the hills. Another picture (**5**) poses an interesting question: did a series of meteors make this extraordinary chain of craters, or was material thrown out from the formation of the large crater? It looks like a stone skipping across water, frozen in time. After 33 hours on the moon, Shepard and Mitchell blast off in the ascent stage of *Antares* at 1:49 PM EST on February 6. They head directly for the CM instead of first going into lunar orbit as did previous missions, and a little later *Kitty Hawk* (**6**) looms out of the celestial void. The docking takes place without any particular incident at 3:35 PM EST.

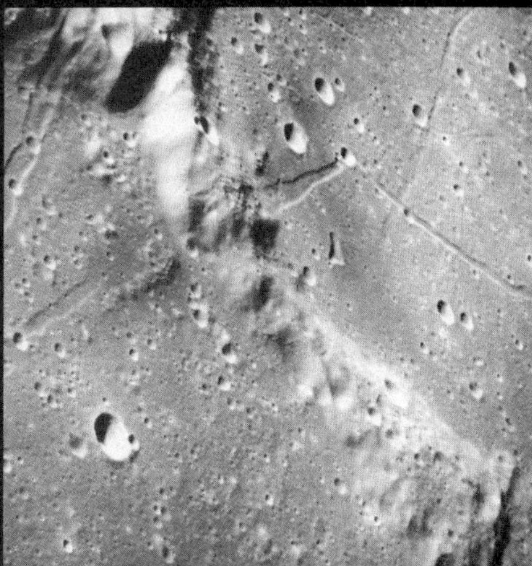

5

APOLLO 14
Splashdown

After Shepard and Mitchell crawl through the connecting tunnel between *Antares* and *Kitty Hawk* for the last time, having carefully transferred the containers with their treasured moon rocks and equipment, *Antares* is jettisoned at 5:48 PM to crash on the moon, which it does almost exactly two hours later. At 8:39 PM the 20,000-pound-thrust SPS engine is fired up once again to push the spacecraft out of moon orbit and into a path back toward earth. The last of their color telecasts from space takes place on Sunday evening, February 7, 1971. After completing a few more experiments, such as testing the casting of metals in weightlessness, electrophoretic separation (a method used to separate biological components in the manufacture of vaccines and medicine), and the transfer of fluids between containers—such as fuels from one tank to another—the astronauts settle into the routine of returning to earth. After three days the Apollo 14 spacecraft arrives in the vicinity of the home planet on Tuesday, February

1

2

3

9, 1971. They jettison the no-longer-needed supply and equipment-carrying Service Module, and the crew, aboard the Command Module, re-enters the earth's atmosphere. A television camera on the deck of the prime recovery ship, the U.S.S. *New Orleans,* catches sight of the spacecraft the moment the drogue chutes open. The three orange-and-white main chutes deploy properly (**1**), and *Kitty Hawk* drops to a safe splashdown in the Pacific Ocean at 4:05 PM EST, 780 nautical miles southeast of Samoa. A minor hitch mars the otherwise perfect splashdown. The big parachutes are rigged to jettison the moment the spacecraft touches the water, to prevent them from dragging the CM away or even overturning it. The shroudline of one of the chutes fails to release; it is cut by one of these Navy pararescuemen (**2**), jumping into the ocean with a life raft from a low-hovering Sikorsky HSS-2. Another pararescueman attaches and inflates the flotation collar (**3**) around the seared spacecraft, and then greets Ed Mitchell (**4**) when the hatch is opened. Conversation is difficult with the germicidal masks the astronauts are wearing, but the joy of homecoming needs few words. The astronauts are carried up one at a time into the helicopter, and shortly afterwards are deposited safely on the flight deck of the *New Orleans.* The welcome of heroes awaits the successful explorers-cum-athletes as they return again to their native planet! Now, after the inevitable fears and doubts cast by the last mission, confidence returns.

4

APOLLO 14
Post-flight

Although splashdown ceremonies started spontaneously at Mission Control, they have now become a ritual performed with nearly the same precision as a space mission. The moment the television camera in the control room shows the Apollo 14 astronauts safely aboard the *New Orleans,* boxes of cigars are passed around, and delighted officials and flight controllers begin blowing smoke rings. Before the crucial re-entry even began, workmen moved quietly through the control room with a ladder, placed it against the control room wall, and hung the Apollo 14 emblem on a hook that had been attached before the mission began, and next to the emblems of previous Apollo flights. Flight controllers in the Mission Control Room (**1**) view the recovery operations on the big monitor screen amid lavish welcoming displays. From the moon to a trailer (**2**) Shepard leads the way as the astronauts enter the Mobile Quarantine Facility isolation van aboard the *New Orleans* after their recovery. Its work done, (**3**), *Kitty Hawk* is

1

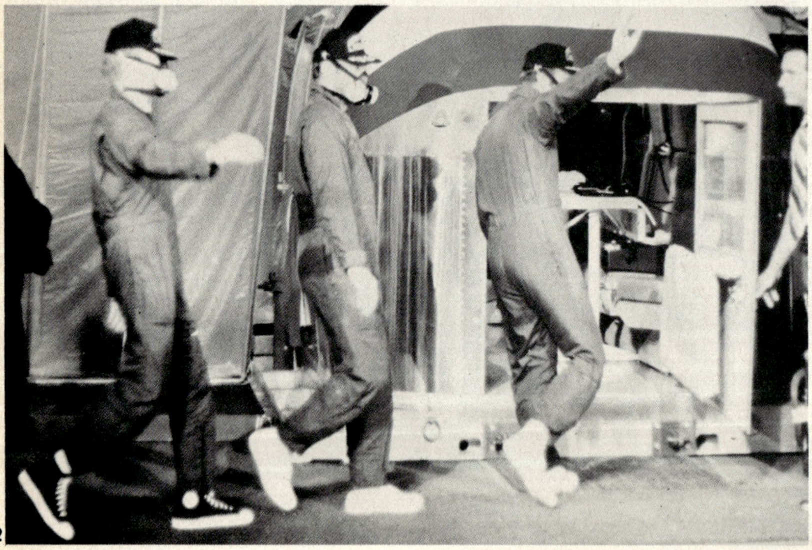

2

3

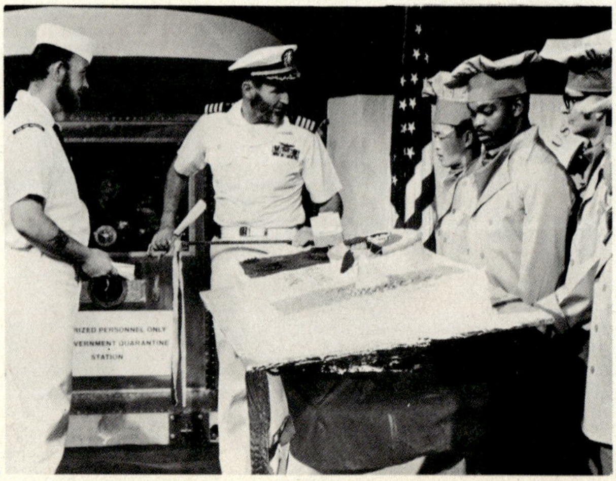

4

5

hoisted aboard. The very image of a grizzled Navy salt (**4**), Capt. Walter Robert Moore of the *New Orleans* presides over the ship's welcoming ceremonies, which include the presentation of a big "Welcome Home" cake by the ship's bakers. Houston was never like this (**5**); the Samoans greet their guests in the native manner as the astronauts arrive on February 11, 1971. They love it. A bit more formal are the ceremonies in Washington, D.C., on March 1, (**6**), where President Nixon awards the NASA Distinguished Service Medal to the astronauts and then assures them and the assembled guests that funds will continue for the space program. Flanked by ranking members of the Senate, (**7**) the astronauts pose on the Capitol steps the following day; Majority Leader Mike Mansfield (left) points out the sights. Ed Mitchell makes a short speech (**8**) about the Apollo 14 mission and the continuing need for space exploration to the assembled members of the U.S. House of Representatives on March 2. Mission completed, the scientists take over, as Leonard E. Cornitius (**9**), lab technician for Brown & Root/Northrop, examines lunar sample No. 14321, part of the 95 pounds of geological specimens brought back by Apollo 14. Nicknamed "Big Bertha", and almost the size of a basketball, this 19.84 lb. rock is the largest sample ever to be collected on the lunar surface by teams of NASA astronauts. It is a coarse-grained brecchia and clastic rock measuring 9.05 by 9.05 by 6.69 inches, and is seen here in the Nitrogen Processing Line.

7

8

6

9

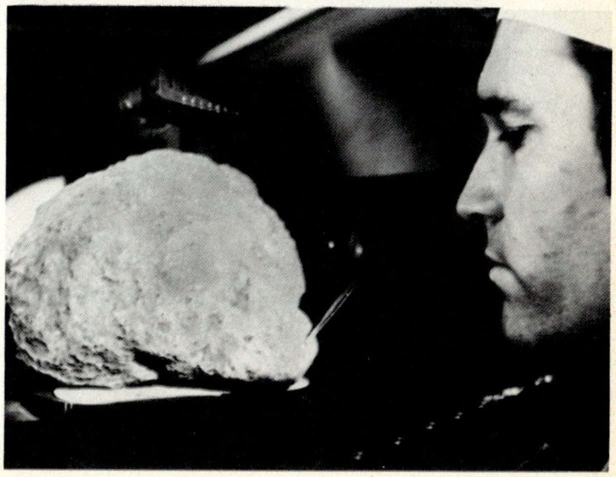

Prior to Apollo 15, astronauts were limited to EVAs within a small radius of the LM. The Lunar Roving Vehicle (LRV) was designed to extend the range of exploration on the lunar surface. Uncomplicated and quite stark in overall appearance (**1**), the LRV is a very sophisticated piece of equipment. Designed and built by Boeing, the Rover has a wheelbase of 90 inches, an overall length of 122 inches, and a height of 45 inches. It weighs only 470 pounds, yet can carry a weight of 970 pounds. Power originates in two 36-volt silver-zinc batteries that are connected to a sealed ½-horsepower electric motor in each wheel hub, giving the vehicle 4-wheel drive. All four wheels also steer, but the front wheels can be disconnected from the rear in case of damage. With a top speed of 10 mph, total range is about 45 miles. The Rover's tires are actually a woven mesh of .033-inch piano wire, with treads of titanium strips rivetted to the mesh. Because of the lunar dust, four fiberglass fenders were fitted. A series of paintings by NASA (**2,3,4,5**) de-

THE LUNAR ROVING VEHICLE

A moon buggy helps to extend man's exploration

2

COURTESY GRUMMAN CO

1

picts the deployment of the LRV after landing on the lunar surface. A special compartment in the LM's descent stage contains the LRV, its frame and wheels collapsed for stowage. While being lowered from its compartment by cables the Rover automatically unfolds, and then the astronauts complete the preparation. Payload arrangement and equipment installation (**6**) includes side-by-side seating for the astronauts. The most evident attachments are the TV camera and its high-gain antenna (**7**), seen here undergoing tests. Mission Control in Houston can operate this camera via remote control and not only monitor the astronauts, but view the lunar landscape in any direction. Much of the available space on board is for carrying ALSEP tools and specimen stowage. The control panel is positioned between the seats and can be operated by either man. Behind the seats is the umbilical link, which ensures a cross-connect between the suits' life support systems should one be damaged during EVA operations. Mounted all the way forward is the Lunar Communications Relay Unit (LCRU), housed in a pack between the front wheels. Tools are stowed aft of the seats, except for the pickup tongs which are conveniently located next to the left seat. Here (**8**) is one of the lunar models in checkout. The technicians in blue are wearing lunar gloves to simulate operation of the controls by space-suited astronauts. Adoption of the LRV for transporting men on the moon was preceded by exhaustive studies and numerous proposals for lunar vehicles.

4

5

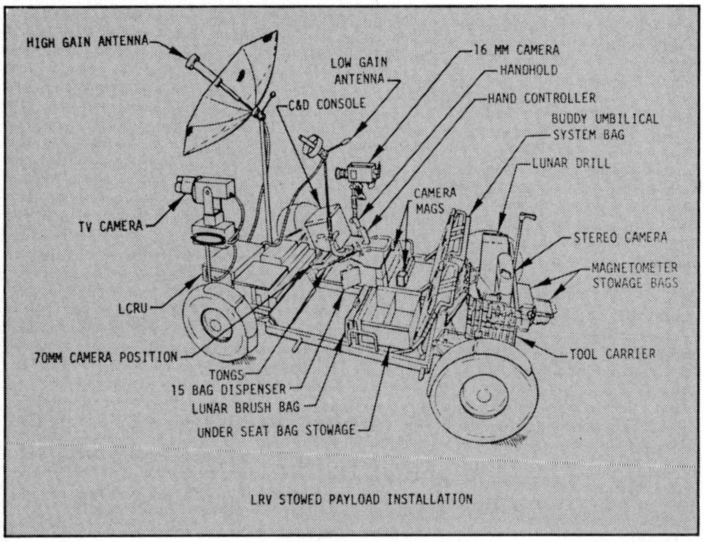

LRV STOWED PAYLOAD INSTALLATION

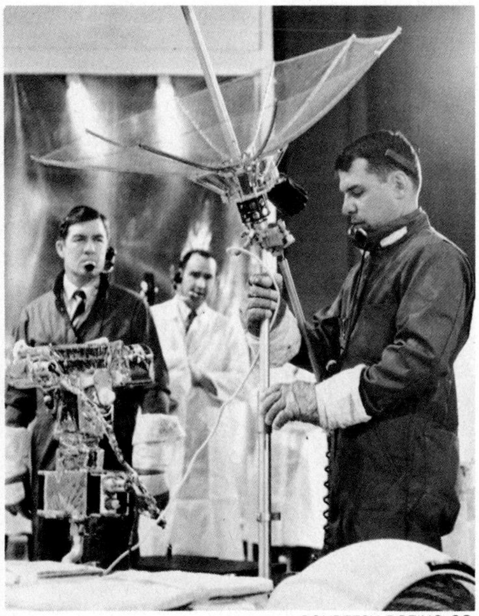

8

COURTESY BOEING CO.

COURTESY BOEING CO.

LUNAR ROVING VEHICLE

In 1970, the Manned Spacecraft Center in Houston conducted a number of tests with two-wheeled vehicles (**1**), under simulated lunar gravity conditions. Undertaken at Building 29, the tests examined feasibility of an astronaut, in complete EVA suit with PLSS, maneuvering about the lunar surface in such a vehicle. Results of these tests led to the development of the two-man LRV. The Rover was envisioned as operating on the lunar surface as a separate link in the communications system via the LCRU pack (**2**). This Lunar Communications Relay Unit acted as a portable station for TV, voice, and telemetry directly between the astronauts and Mission Control, instead of the LM. Not including antennas, the LCRU measured 22 x 16 x 6 inches, and weighed 55 pounds on Earth, but only 9.2 on the lunar surface. It contains its own power pack and folding antenna, and can be carried by a crewman, or stowed at the forward end of the LRV. A protective thermal blanket (**3**) may be drawn back to vary the total radia-

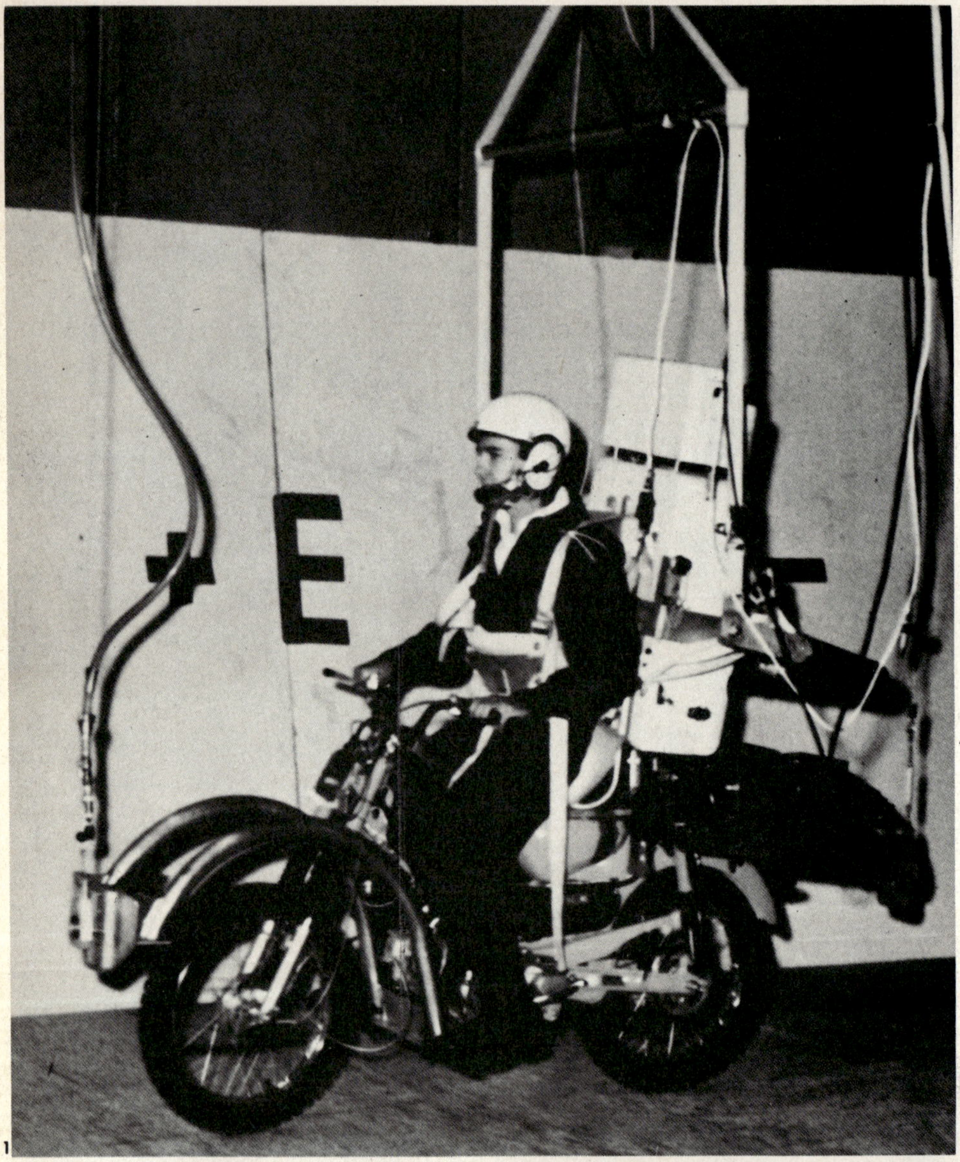

1

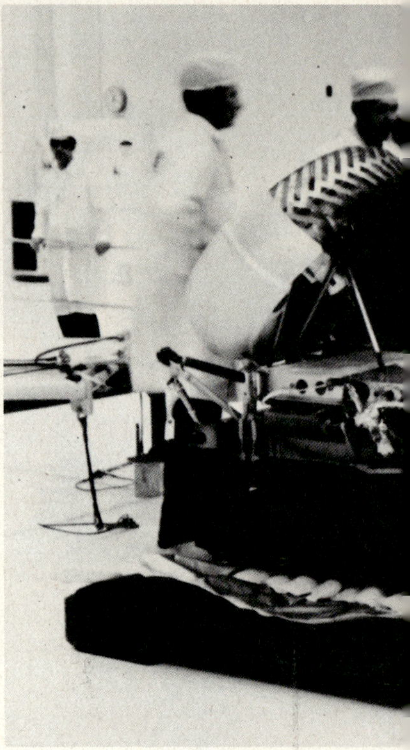

4

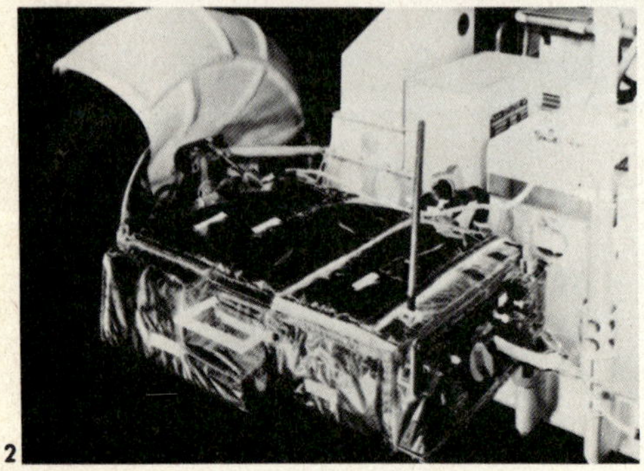

2

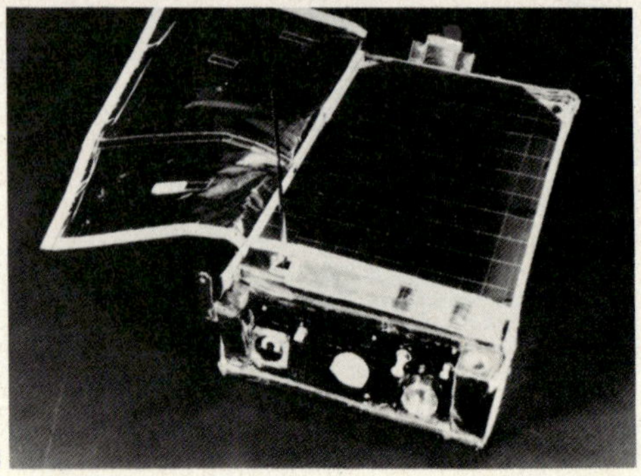

3

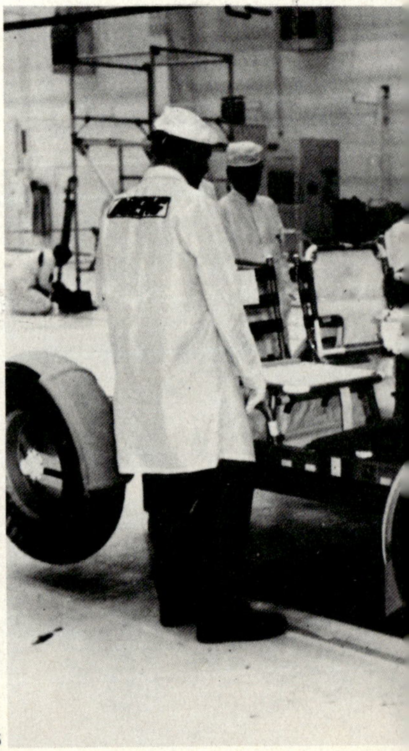

5

tion surface. In addition, 196 square inches of radiating mirrors reflected solar heat. Thermal stability in the LCRU is ensured by a ¾-inch layer of wax on its upper surface which underwent melt/freeze cycles on the moon. One of the most interesting problems faced by Boeing engineers and technicians was the in-flight stowage of the Rover. The answer (**4**), resulted in a compact, roof-shaped package which could be stowed in a special compartment on the LM. The frame and wheel units are designed to collapse and fold into a package 66 inches in length and 36 inches high. It can be seen that the wheels are folded inwards, then the frame ends to which they are mounted are "flopped" over towards the center control console. Technicians simulating deployment (**5**) give an excellent idea of the LRV's small dimensions. Note the seats equipped with mandatory seat belts. The control and display console (**6**) is split into two functions. The upper section contains navigational components, the lower houses control switches for the electrical drive control. The "joy stick" is located centrally and allows one-handed steering and braking action. Pulling rearward on the stick not only reduces current, but actuates cable-controlled brakes. The thumb switch on the stick activates the reverse mode. Under a 38.5-million-dollar contract Boeing built four lunar-ready Rovers (one each for Apollo missions 15, 16, 17 and a spare) and several test vehicles, including a full-scale trainer with rubber tires for simulated lunar terrain.

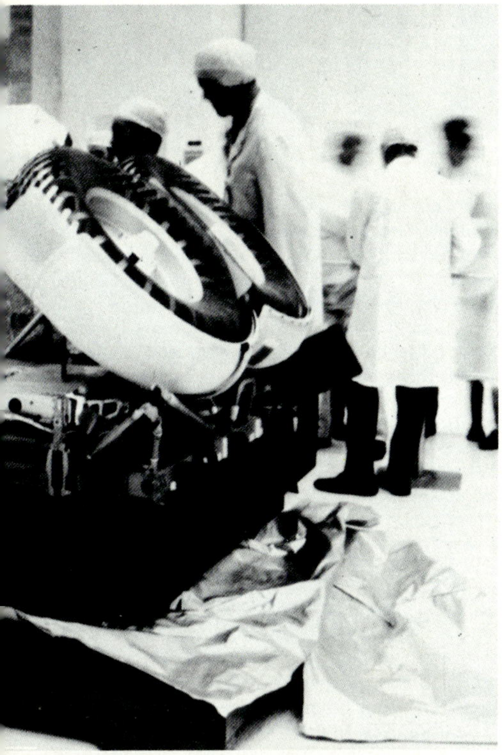

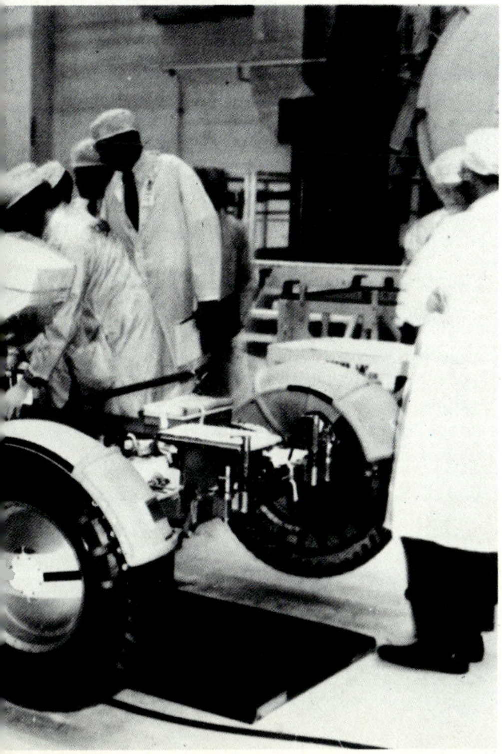

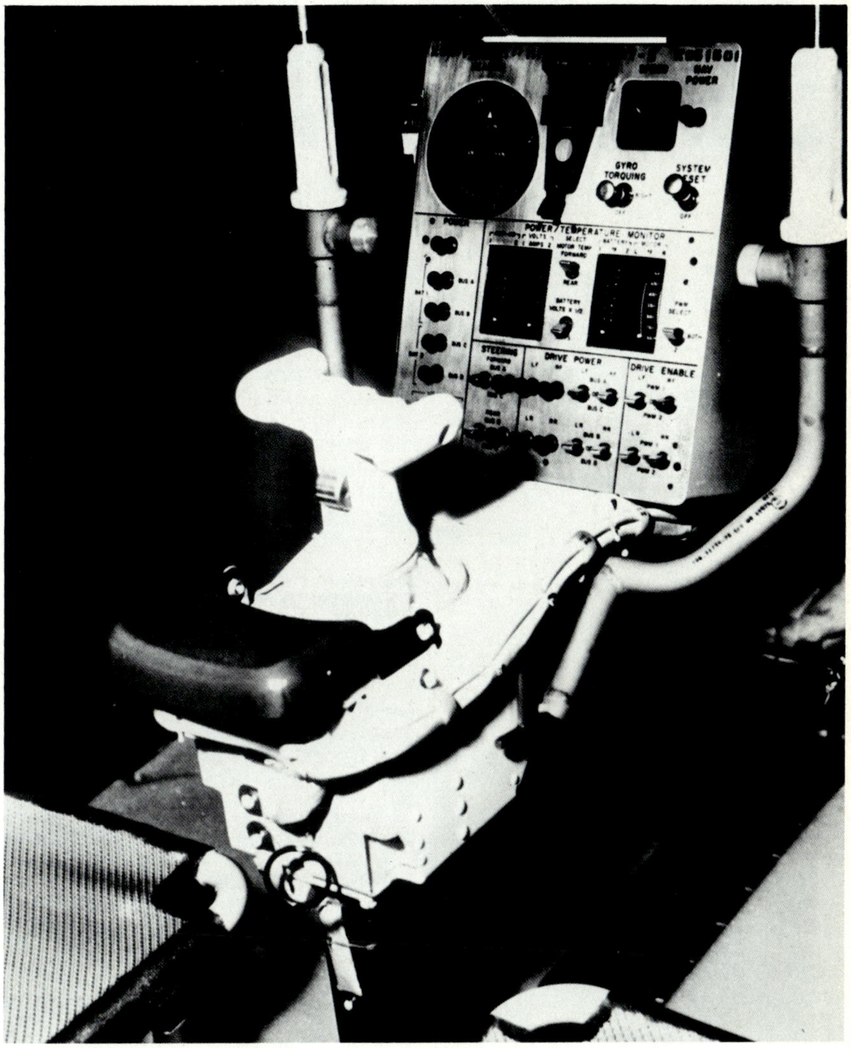

6

APOLLO 15

Visualize: the gently rolling (but crater-scarred) surface of the Marsh of Decay. It is 6:15 PM, July 30, 1971, EDT, five days into space. The Lunar Module *Falcon*—named for the USAF mascot—gingerly touches the moon with its four, 37-inch circular foot pads. Mission Commander David R. Scott reports to Houston that, very simply, *"Falcon* is on the plain at Hadley." And indeed she is, in spite of Scott's following message: "We're on a ten degree slope." Apollo 15 is the fourth manned lunar landing mission and the first to visit and explore the moon's Hadley Rille and Apennine Mountains, on the edge of the Sea of Rains. Subsequently, a new record was set for surface stay time—66 hours and 54 minutes. The mission demonstrates man's increasingly audacious lunar ego. With Lunar Roving Vehicle-1, Scott and lunar module pilot James B. Irwin will log 18 hours and 35 minutes each in extravehicular activities, during three separate lunar excursions. Apollo 15 will achieve impressive firsts, including: nearly 180 pounds of lunar soil, largest payload placed into earth and lunar orbits, first subsatellite launched from lunar orbit, and the first deep-space extravehicular activity from a Command Module during the transearth coast.

DAVID R. SCOTT

SPACECRAFT COMMANDER/ A veteran of Gemini 8 and Apollo 9 missions, USAF Colonel Scott is a familiar figure at Houston and Cape Kennedy; he also served as backup Commander for Apollo 12. A West Point graduate, he holds Bachelor of Engineering and Master of Science degrees in Aeronautics and Astronautics from MIT. A much-decorated pilot by both NASA and the Air Force, Scott has more than 5000 hours in the air. One of his hobbies is photography, and a trip to the moon just has to be any photo bug's dream.

ALFRED M. WORDEN

COMMAND MODULE PILOT/ Among the 19 astronauts selected by NASA in April, 1966, Lt. Colonel Worden entered the USAF following graduation from West Point in 1955. He has earned MS degrees in both Aeronautical/Astronautical Engineering and Instrumentation Engineering from the University of Michigan, and logged more than 3600 hours flying time, including 2900 in jet aircraft. Never in space before, he was backup CM pilot for Apollo 12. Born in Jackson, Michigan, February 7, 1932, he enjoys skiing and swimming.

JAMES B. IRWIN

LUNAR MODULE PILOT/ Selected by NASA in 1966, Jim Irwin was born in Pittsburgh on March 17, 1930. He graduated from the Naval Academy in 1951, but entered the ranks of the Air Force, now holding the rank of Colonel. He earned MS degrees from the University of Michigan in Aeronautical Engineering and Instrumentation Engineering. Formerly Chief of the Advanced Requirements Branch, Air Defense Command, he has logged over 7000 hours air time, including time as a test pilot. He was the backup LM pilot for Apollo 12.

APOLLO 15
SCOTT WORDEN IRWIN

APOLLO 15 Preparation

An up-coming lonely vigil in lunar orbit of 67 hours faces Apollo 15 Command Module pilot Alfred M. Worden. At the Kennedy Space Center (**1**), Worden practices alone in a mission simulator. His "upstairs" time will see exploration by the first Lunar Roving Vehicle (LRV), containing astronauts David R. Scott, Mission Commander, and Lunar Module pilot James B. Irwin. (**2**) Familiarization with the LRV comes early for Irwin (left) and Scott, during a pause in testing. The LRV has passed a fit check (**3**) at the Manned Spacecraft Operations low bay area. The LRV qualification team is pleased with the fit of the folded moon car inside the Lunar Module. Note the large quantity of big inspection tags. A night view of Launch Complex 39A shows searchlights trained on the Saturn V space vehicle (**4**), as technicians install Apollo 15's instrument packages. Among the key pieces of gear is the panoramic camera (**5**), designed to obtain 3-dimensional, high resolution ground imagery of potential lunar landing sites. (**6**) The

1
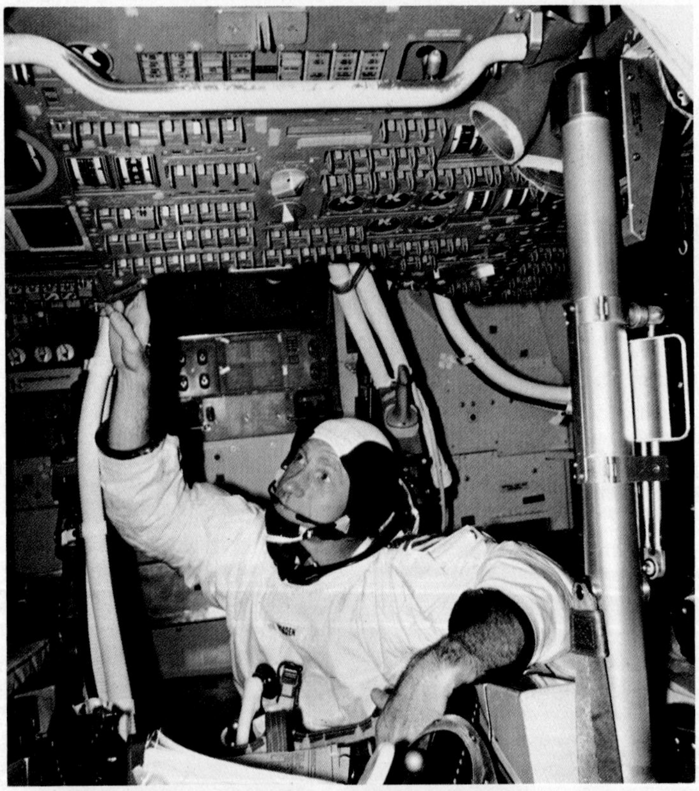

2
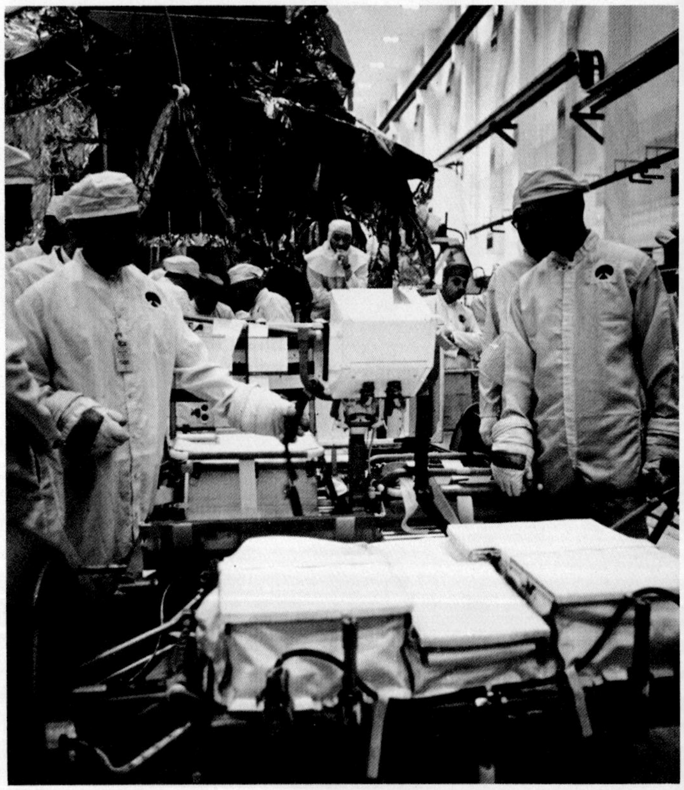

5

6
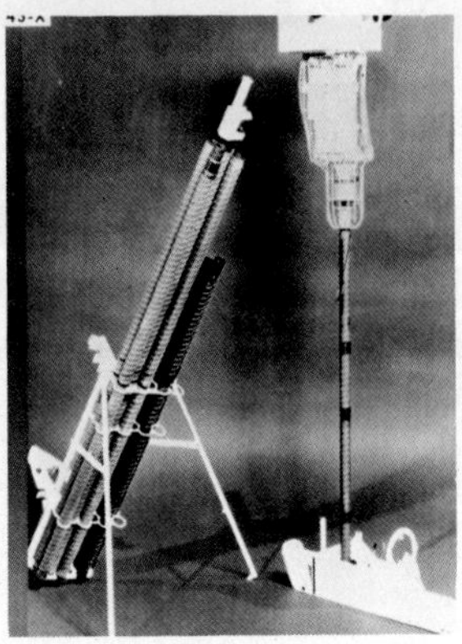

7

8

moon's surface will yield more secrets with the Apollo Lunar Surface Drill (ALSD)—a hand-held, battery-powered, rotary-percussion drill. The objective of the ALSD is to allow an astronaut to emplace Heat Flow Experiment (HFE) probes below the lunar surface and to collect subsurface cores. The HFE probes require boring two subsurface holes to a maximum depth of 3 meters, and will allow soil to be carried back to earth in an Apollo Lunar Sample Return Container. Not a drill, but sending powerful beams of a different kind (**7**), is the Laser Ranging Retro-Reflector, deployment of which will allow for a triangular laser reflector network on the moon. The instrument contains 300 corner reflectors, three times the number contained with similar devices employed in Apollo 11 and 14 missions. Another device is the jettisonable 80-pound sub-satellite (**8**) being stowed by engineers, and designed to study interaction of the earth's magnetic field with the moon. Not forgotten, though, is the precious human element, as technician Rita Rapp (**9**) packages the astronauts' space food. A pint-sized power (**10**) is the Hasselbland 70mm lunar data camera. Note large tabs on setting rings behind the lens for use with lunar gloves. (**11**) Picture is "OK for go" for Apollo 15's Lunar Module *Falcon* (LM10) following altitude checkout. And high atop Launch Complex 39A (**12**), engineers Ray Malone (left) and Ann Montgomery examine stowable gear. The next element, the capricious weather, sets fears that the launch will probably be delayed.

4

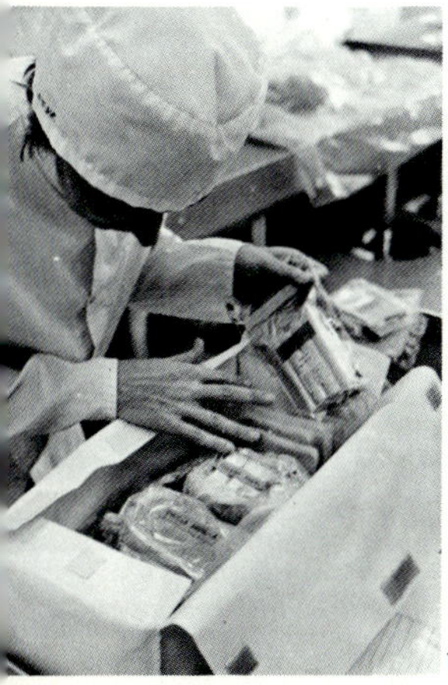

10

11

12

APOLLO 15
Launch

The month of June 1971 sees rain and foul weather in the Cape Kennedy vicinity. Apollo 15's service tower feels lightning strikes at least 10 times before the launch day dawns, July 26th, under almost cloudless skys. There were some who thought that public interest in space exploration had waned by early 1971, but Apollo 15's ambitious goals and perhaps the all-Air Force crew, did nothing to dampen an overwhelming public turnout. Lift-off day for Apollo 15 brings the largest crowd since Apollo 11. Major public highways adjacent to the Cape are jammed (**1**), while camping and parking spaces are at a premium. The launch is a typical procedure, demanding thorough planning—of which this portion of the space vehicle Countdown Demonstration Test is an example (**2**). The "dry" run, held a week earlier, shows Mission Commander Scott (left) speaking with Donald K. Slayton, director of Flight Crew Operations. Worden (center, rear) and Irwin (seated, rear left) also participate. Today, Mission Commander Scott

1

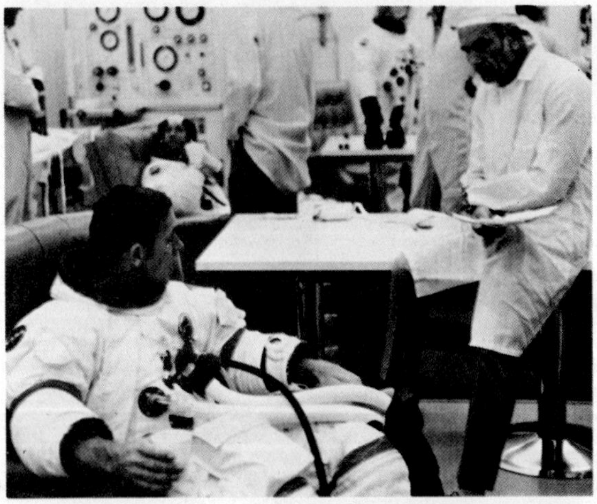
2

3

4

leads astronauts Irwin and Worden (**3**) from the Manned Spacecraft Operations Building to the transfer van at the start of the 9-mile trip to the launch pad, while a camera atop the Launch Umbilical Tower's 360-foot level (**4**) records the A/S 510 Saturn V launch vehicle venting liquid oxygen. Another Countdown Demonstration Test prior to launch day (**5**), has Scott (front) and Irwin walking across an access arm on the mobile launcher, en route to their spacecraft. Today, however, it's the real thing (**6**), as technicians prepare to help Scott into the spacecraft (*Endeavour* is spacecraft #112). LM pilot Irwin awaits his turn. The countdown is flawless, and at 9:34 AM EDT, three more fragile humans place their faith in man's ability to harness 7½-million pounds of thrust pointed to the heavens. The sequence on following pages 88 and 89 shows the power of the 363-foot-high space vehicle at lift-off. A panoramic camera exposure of 1/1000-second catches water rushing down the Mobile Launcher at 45,000 gallons per minute to contain the rocket's fire and prevent it from destroying the Mobile Launch Platform. Up thunders the space vehicle, through the troposphere, into the stratosphere; Apollo 15 is A-OK. A tracking camera records the first 2½ minutes of flight, and the first stage engine shutdown. The successful launch spreads relief and enthusiasm (**7**) in Firing Room 1 of the Launch Control Center. All the endless preparation, cooperation, precision, "zero defects" and science, has produced a stunning liftoff.

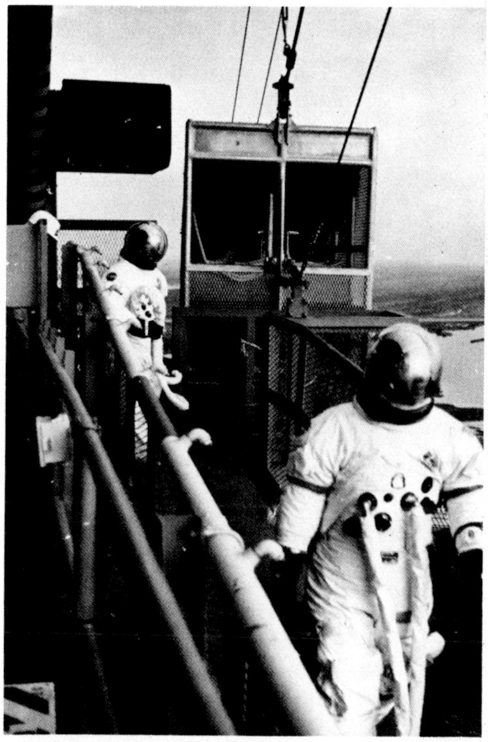

5

6

7

APOLLO 15
Launch

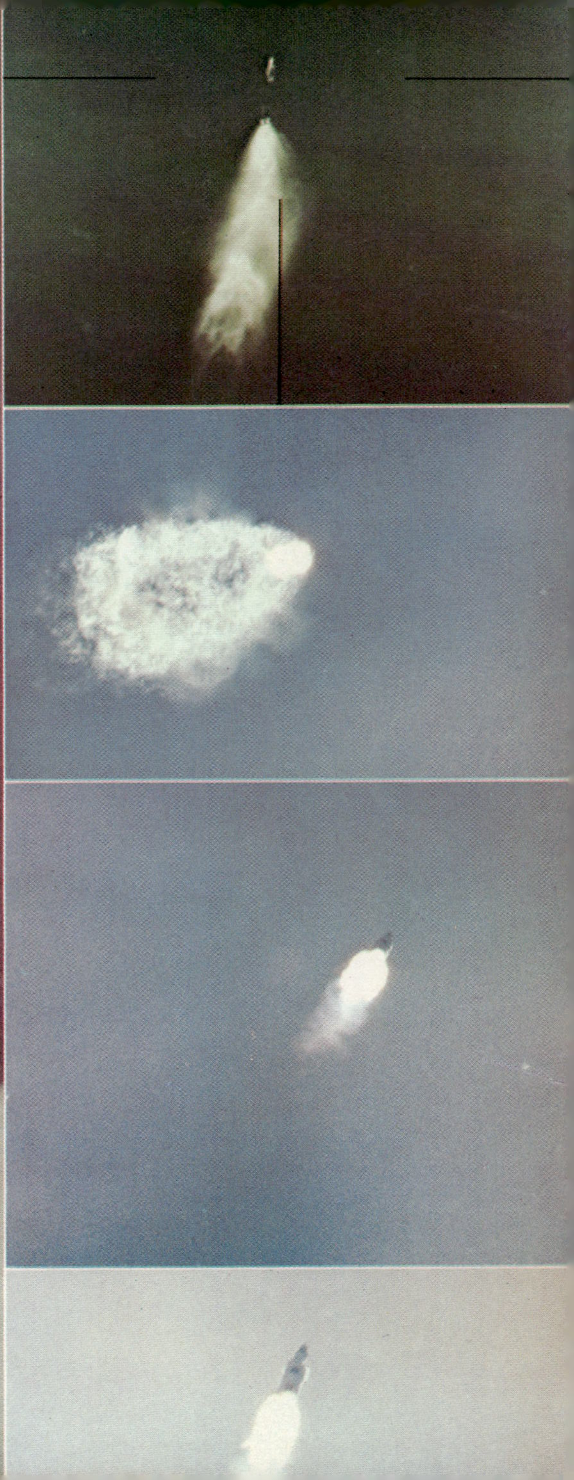

APOLLO 15 Exploration

Two hours and 50 minutes after lift-off, Apollo 15 rockets out of earth orbit toward the moon. Inside the Command Module *Endeavour,* the astronauts look at a map of the lunar rover route. The LRV will participate in three extravehicular activities, as the NASA map indicates (**1**). *Falcon* is to land at the point labeled "site; the crater size key at left references Apollo 14 craters. The landing site is several kilometers east of Hadley Rille, with the Apennine Mountains off the map to the east. Landing site activities will include five objectives; the first and primary objective is to sample the mountain front. The second objective is to study Hadley Rille, the mile-wide, meandering canyon of 600 to 1200 feet in depth. Third and fourth objectives will include observation of *Falcon's* landing site, the Marsh of Decay, and the north complex, the latter expected to be volcanic. The final objective is the cluster of craters southeast of the landing site. On July 29th, three days after lift-off, Apollo 15 sweeps around the moon, fires its engine to

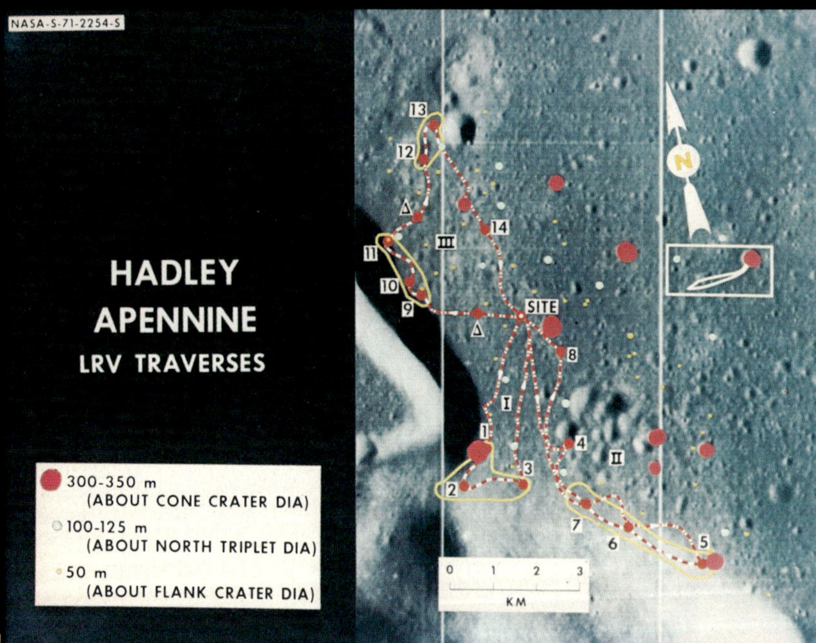

1

2

5

6

slow down, and enters lunar orbit. Other phases, common to Apollo missions, follow in sequence. Touchdown comes at 6:15 PM, EDT at the Hadley-Apennine site. Then, at 106 hours from lift-off, Mission Commander Scott stands up through *Falcon's* hatch to survey the moonscape. One of the early views is of Hadley Delta (**2**), looking southeast toward Silver Spur. Another view of Hadley Delta has astronaut Irwin saluting the flag (**3**) with the camera facing almost due south, and St. George Crater behind the LRV about 5 kilometers away. An astronaut's look shows an undisturbed lunar rock forming a stark contrast to bootprints (**4**). While in lunar orbit, Worden uses his countless hours of preparation to advantage, by photographing lunar surface features from *Endeavour* (**5**). Notice the glass reflection, caused by the glaring sun. At moonbase, a mosaic photograph shows a 360-degree panoramic vista of *Falcon's* landing site. Mount Hadley is the prominent feature at the right center. Scott is reaching to pick up the Apollo Lunar Surface Drill (**6**) while components of the Apollo Lunar Surface Experiments Package (ALSEP) are deployed with various electrical connections extending from the ALSEP Central Station at left. Ever watchful, flight controllers in Mission Control Center (**7**) keep a constant eye on the astronauts and the complex systems of the LM and CSM with their monitors, and the Capcom is always in communication. Even while the crew sleeps, their far-distant guardians in Houston are alert and watchful

4

7

APOLLO 15 Exploration

Falcon's base on the Marsh of Decay enabled the Apollo 15 expedition to study, in a single mission, three types of lunar topography—the mare basin (plain), lunar rille (gorge), and the mountain front. Mount Hadley, fully lighted (**1**), shows abundant linear features. As part of the Apennines, Mount Hadley is believed to be composed of material thrust up from below the surface by tremendous internal pressures built up when an object about the size of Rhode Island collided with the moon and created the Sea of Rains (Mare Imbrium). Mt. Hadley rises approximately 14,765 feet above the plain. Another spectacular sight follows, this time through the camera's magic in a top-to-bottom sequence of *Falcon's* blastoff (2, 3 and 4, reading down). First of the three photos (**2**) shows the Lunar Module only seconds before the ascent stage liftoff, in this reproduction from a color television transmission beamed to Mission Control in Houston. The TV camera was mounted on the LRV, parked about 300 feet east of *Falcon*.

2

3

1

4

5

6

7

The second photo in the series shows a kaleidoscope effect created by the flame from the LM ascent stage during liftoff (**3**), which was recorded at 171 hours and 37 minutes, August 2nd, Ground Elapsed Time. And the bottom photo shows the *Falcon* descent stage, resting alone, shortly after the ascent liftoff. Here (**4**), the lunar soil dust cloud stirred up by blastoff has already settled, and the 66-hour and 55-minute exploration by astronauts Scott and Irwin has established man's longest and most rewarding moon expedition. Looking back, again through the magic of television transmission, Scott was noted kneeling to pick up a lunar sample during the third EVA (**5**); Irwin joining Scott (**6**), also during the third EVA, for more lunar sample probing. Looking like strange apparitions indeed (**7**), with their spacesuits and probing tools, Scott (left) and Irwin—seen from EVA 2's TV ground image—team up to take soil penetrometer tests. The object of such tests is to determine physical characteristics and properties of the lunar surface. As *Falcon* heads for docking with CM *Endeavour* in lunar orbit, it was also remembered that one of the more spectacular photos (**8**) caught lengthening shadows of the moonscape, with Irwin at the LRV. The lunar rover is parked near Falcon, looking northeast, with Mount Hadley in the background. The Lunar Roving Vehicle, descent stage of the LM, and other objects remain forever on the surface. Summing up this mission, Scott said, "Man must explore, and this is exploration at its greatest."

APOLLO 15
Splashdown

Here (**1**) is *Endeavor* at rendezvous, seen from the LM as the two craft maneuver for docking. An opposite view, of *Falcon* (**2**) shows the ascent stage approaching *Endeavour*. After docking in lunar orbit, another dramatic moment is prepared, from *Endeavour's* Scientific Instrument Module (SIM) bay. From a space about 9½ feet high by 5 feet wide, the SIM bay instrument packages—visible with their access door jettisoned— have been recording and photographing large areas of the moon's surface. Now from the SIM bay is launched a 78.5-pound scientific subsatellite (**3**), to gather about a year's data on lunar gravity, mascons, solar flares and magnetic fields of both the earth and the moon. An hour after launch of the subsatellite, *Endeavour* fires its engine to accelerate out of lunar orbit toward earth. The next day, August 5th, Worden climbs from the CM and retrieves two cassettes of film from the SIM bay (**4**). Worden's space-walk occurred some 171,000 miles from earth, at a CM speed of

1

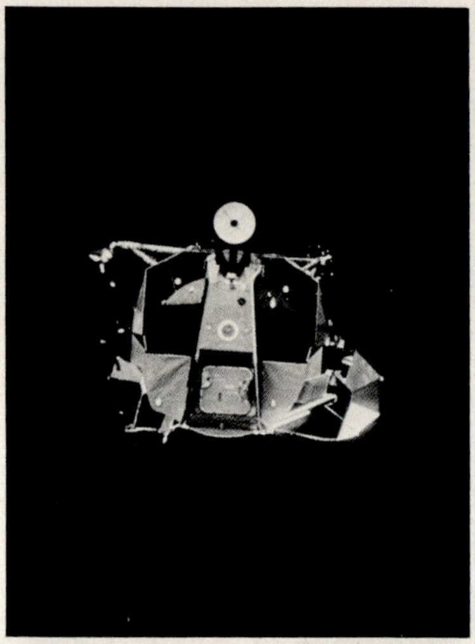
2

5

3

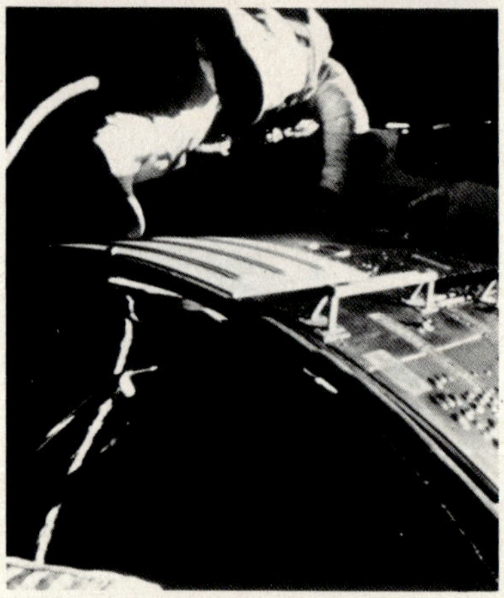
4

6

8

9

10

about 3,000 miles an hour. This is the longest walk yet in deep space, lasting about 16 minutes. Homeward speeds *Endeavour,* on its 3-day flight with everything A-OK. Retrofire braking, Service Module jettison, and re-entry proceed on schedule and without incident. Following the plunge through the earth's atmosphere the chutes blossom forth. Seconds before splashdown, one of the spacecraft's three main descent parachutes collapses (**5**). The result causes an increase in impact velocity from 19 to 22 mph, but the additional velocity doesn't jeopardize the safety of either the spacecraft or the astronauts. Splashdown comes at 4:46 EDT August 7, 1971, 333 miles north of Hawaii. Minutes later, the scorched spacecraft is cradled inside a flotation collar (**6**) as the USS *Okinawa* rides easily nearby. The last to leave *Endeavour* is Mission Commander Scott, assisted by a US Navy frogman (**7**). Worden, Irwin and Scott (from left) await helicopter pickup (**8**), and 40 minutes after splashdown, they are whisked aboard *Okinawa.* Both the prime recovery ship and the crew of Apollo 15 have now accomplished their missions, as *Okinawa* heads east (**9**). Later, a special card is presented to Chicago's Mayor Daley at city welcoming ceremonies (**10**). A guest appearance on the Dick Cavett TV show follows (**11**). Apollo 15 set a new record for lunar surface stay time—66 hours, 54 minutes—a new record for lunar EVA time—18 hours, 35 minutes—and brought back nearly 180 pounds of samples, including this basalt rock (**12**).

7

11

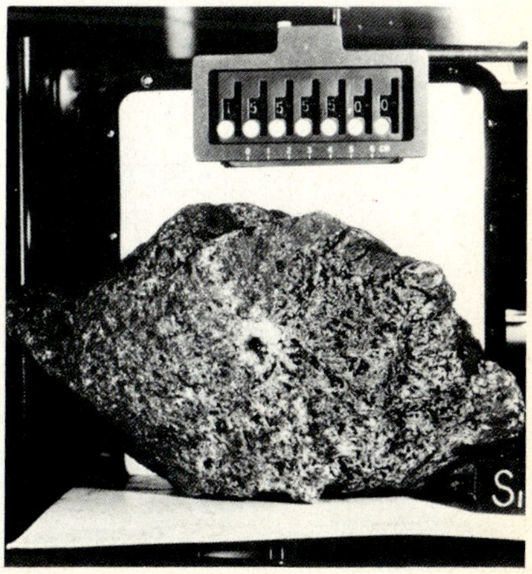

12

During the lunar phase of the Apollo missions, the Command/Service Module remains in orbit around the moon, and it plays "second fiddle" to the unfolding drama on the surface. Prior to Apollo 15, the LM Pilot's duties in orbit consisted primarily of photomapping and surveying future landing sites. Aside from these, he was responsible for his own well being and that of the CSM's, but there were many lonely hours in silent orbit. The routine has been entirely revised for Apollos 15, 16 and 17. For these missions, the CSM includes a complex pack of highly sophisticated scientific equipment, located in the Service Module. NASA refers to this pack as the Scientific Instrument Module Bay (SIM Bay) and the entire mission role as the type "J." Thus, the J missions enable the CSM to become an advanced orbital laboratory under the guidance of the CM Pilot. Photographed from the LM (**1**) is Apollo 15's SIM bay. Protective doors are blown off by explosive bolts prior to deployment of the instruments. A

THE SCIENTIFIC INSTRUMENT MODULE BAY

Scientific Instrument Module lets CM Pilot map the moon

1

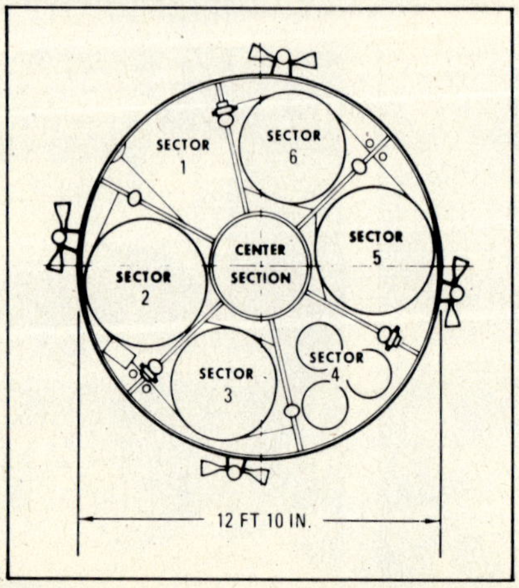

2

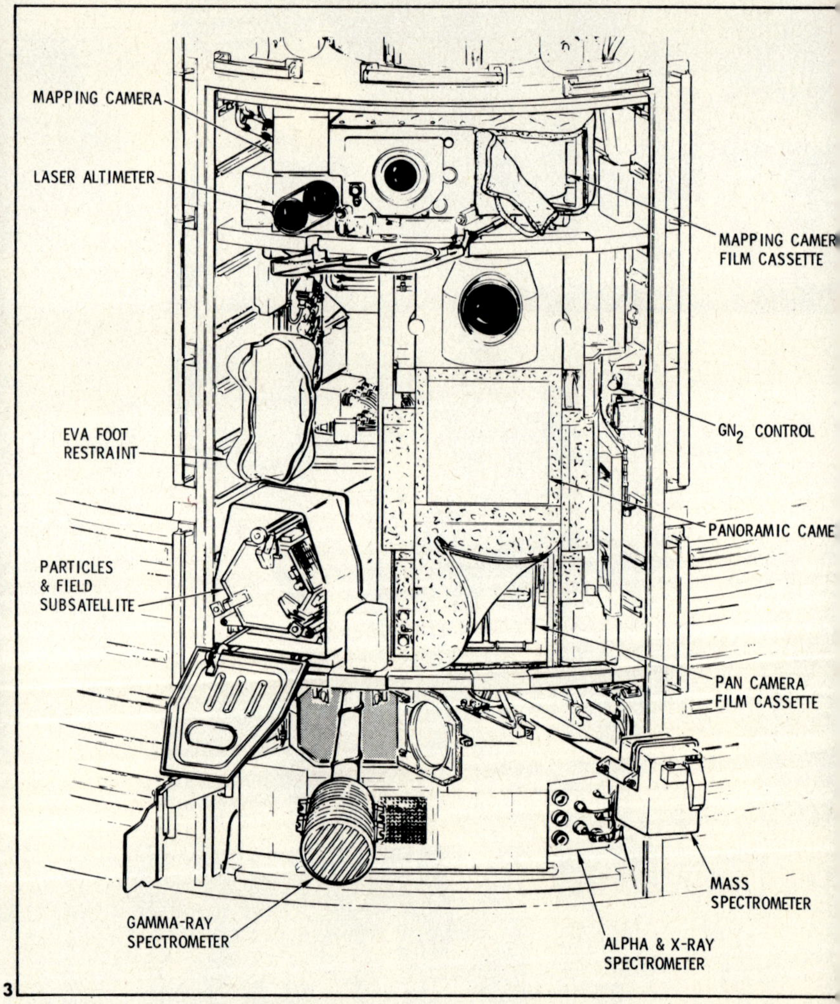

3

cutaway drawing (**2**) of the SM's center section reveals the radial location of the SIM bay in Sector 1. During operation, this sector faces downward toward the lunar surface. North American Rockwell, prime contractor for the SIM bay, provided this layout drawing (**3**) of the Apollo 15 bay (each of the three differs). Each of the individual pieces of equipment is identified. Apollo 15's equipment includes a small satellite to be launched during orbit. The tiny 80-pounder, projected into an elliptical orbit by a spring-loaded mechanism, will remain in orbit up to a year and transmit data about the solar energy field around the moon. This painting (**4**) shows the Apollo 16 bay deployed. Long probes are the gamma ray and mass spectrometers. The mass spectrometer will measure any lunar atmosphere, while the gamma ray unit will measure gamma radiation from the surface—an indicator of the moon's true age. (**5**) The launch of the subsatellite from Apollo 15 to study the magnetic particle field surrounding the moon. During the return (**6**) trans-earth coast, the CM Pilot carries out an EVA, retrieving the film and assimilated data tape cassettes from the SIM bay. He is observed and aided by the LM Pilot. A TV camera records the events. The CM Pilot is restrained by a tether and foot restraints. An emergency oxygen supply is carried by him. The LM Pilot and Commander don their lunar suits for this operation, as the CM is depressurized. Handrails surround SIM bay which allow the CM Pilot to pull himself along the hull of the Service Module.

4

5

6

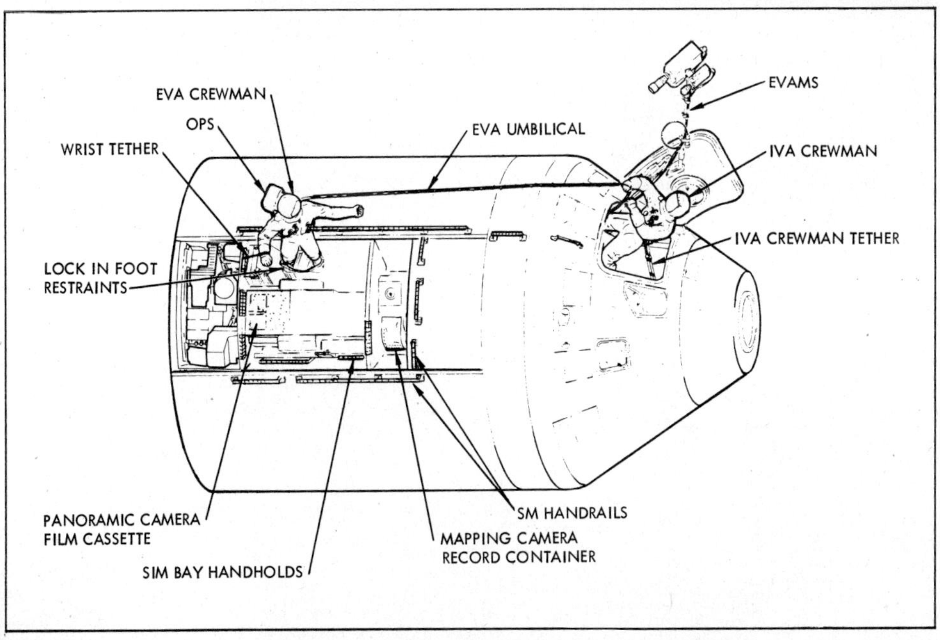

7

A rather delicate subject? Well, to a man jammed into the confined interior of a space ship, it is not a delicate subject; it's an operational necessity. The need for toilet facilities became apparent as soon as extended orbital capacity was reached with the Mercury capsules. At first, NASA borrowed heavily from aircraft techniques. Fighter aircraft on long over-water flights are equipped with a simple tube and funnel arrangement in the cockpit, known universally as the "relief tube." In space, however, liquids present a real problem. Water forms spherical droplets and will float around the cabin where it can get into electrical panels and air-conditioning systems, causing short circuits. As a result, liquids—be they water, coffee, or urine—must be contained. Water and other drinkable liquids are stored in flexible bags equipped with a short neoprene tube. An astronaut's lips provide the seal during the liquid transfer process. With urine, an astronaut's bladder is the flexible container, and the seal for the liquid transfer process is provided by a rolled elastic cuff attached to a flexible hose leading to a temporary storage tank. For Mercury and Gemini missions, the urine system was relatively straightforward (**2**), with individual bladders and larger storage bags. The Apollo system is more complex, as this drawing (**3**) of the components in the Command Module shows. Each

WHERE'S THE BATHROOM?

An answer to the question TV commentators never asked

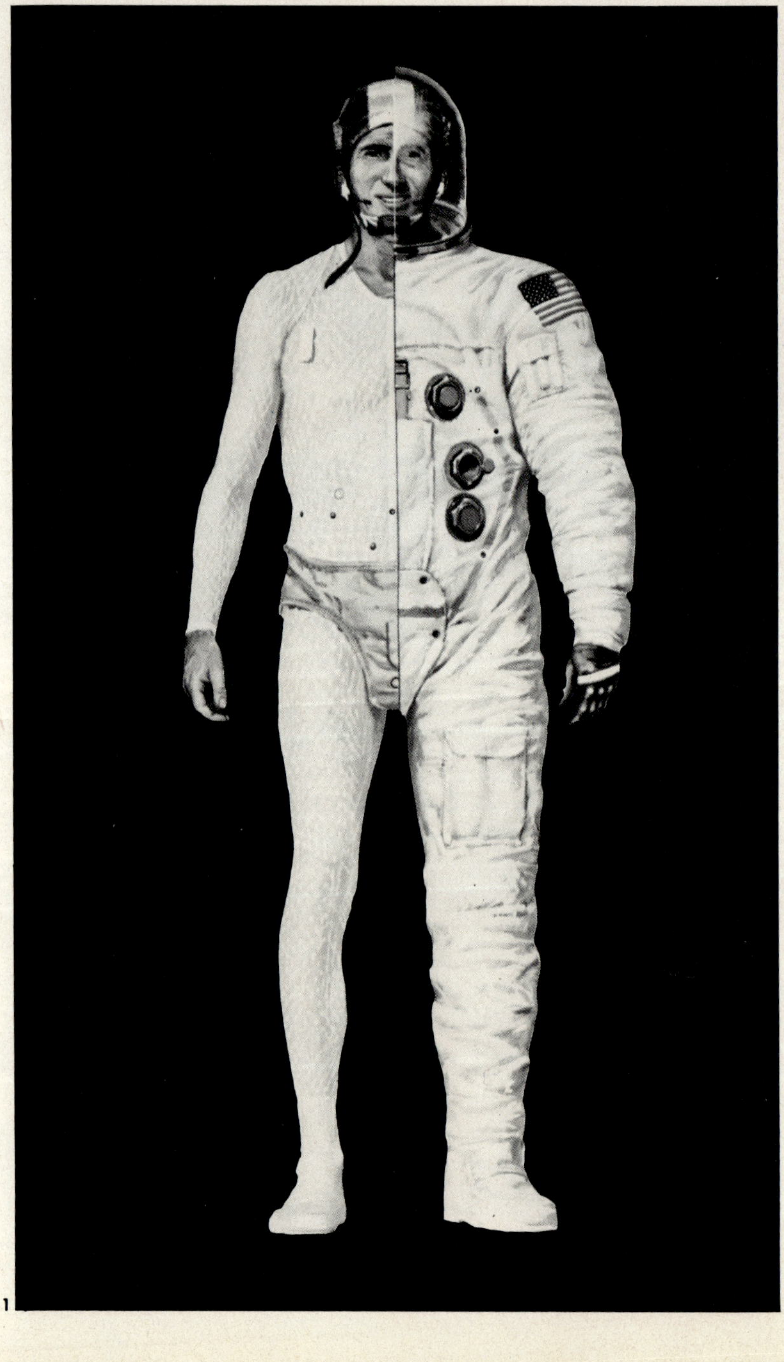
1

2

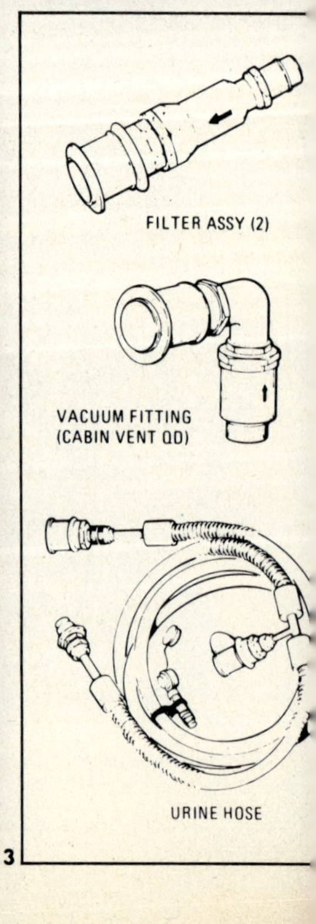

3

astronaut has his own diaphragm and roll-on cuff, kept stowed when not in use. The diaphragm fits into a urine receptacle assembly, which transfers urine to a large storage bag through a hose long enough to reach any of the three couches. If the astronaut is wearing a spacesuit, underneath it is worn a storage bladder (**1**), which is later emptied into the storage bag by means of the urine transfer system. From there it is vented into space through the cabin vent vacuum fitting, along with contaminated water from the batteries and air conditioning system. The Lunar Module has a similar system. Fecal containment presented a different challenge. Beginning with orbital Mercury missions, the astronauts were put on a high-protein diet a few days before the flight, thus cutting down on the need for bowel movements. The same was done with the Gemini astronauts. However, with missions lasting more than a week, the problem had to be faced. Since neither crewman had much freedom of movement in the Gemini capsule, a simple "diaper" was worn (**4**), and the astronauts were soon voicing the same complaints as toddlers the world over; they were wet, uncomfortable, and developing diaper rashes. On Apollo missions lightweight disposable bags are used, feasible because the Command Module and Lunar Module are large enough for limited movement. These bag assemblies consist of two bags, one inside the other, with the inner bag containing the fecal matter. Attached to the inner bag is a pouch containing a germicide. For use, a thin plastic layer is peeled off the inner bag rim, exposing an adhesive backing, and the bag is pasted to the buttocks. After use, the bag is sealed and the germicide pouch emptied into the inner bag, which neutralizes bacteria and prevents outgassing. This latter feature is particularly important in space, since cabin pressure is only five pounds per square inch; pressure from decaying waste could literally explode the sealed bags. The used bags are stored in the special waste disposal compartments.

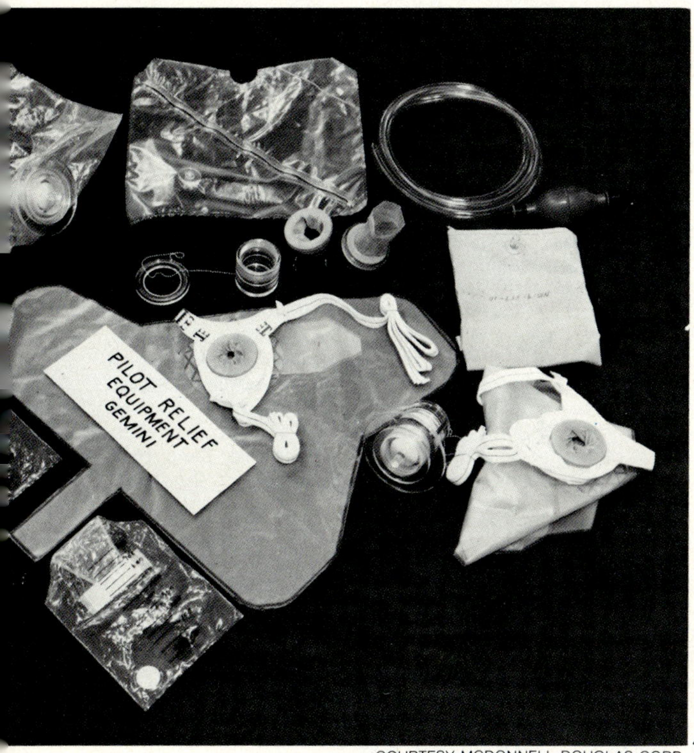

4

COURTESY MCDONNELL DOUGLAS CORP.

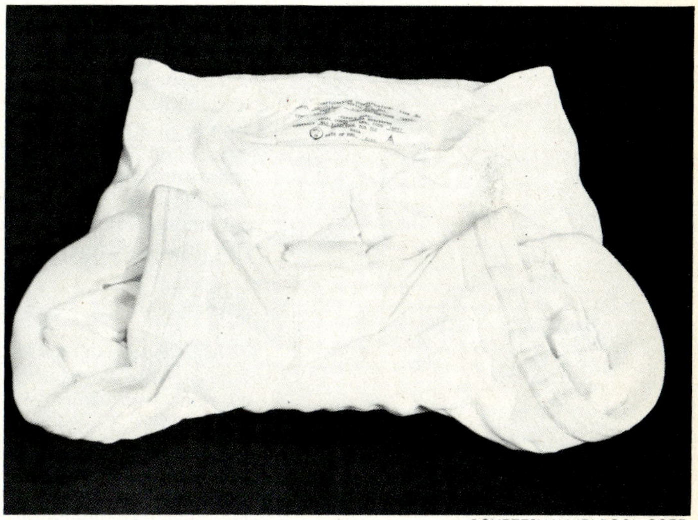

COURTESY WHIRLPOOL CORP.

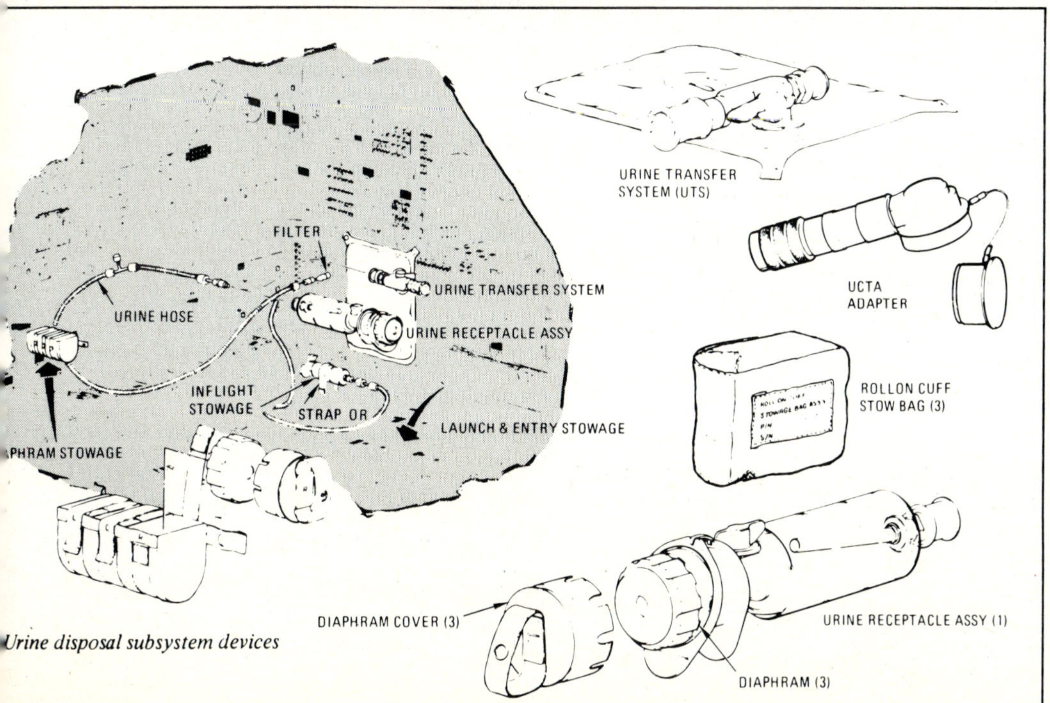

Urine disposal subsystem devices

COURTESY ROCKWELL INTERNATIONAL

APOLLO 16

Apollo 16, the fifth of man's expeditions to the moon, is by far the most sophisticated. Since Apollo 11, our capabilities for conducting intensive scientific investigation in the moon's totally hostile environment have increased substantially. The Lunar Roving Vehicle has increased surface mobility, and the scientific payload on Apollo 16 is 1050 pounds, quite a jump over Apollo 11's 250 pounds. Another "first" of extreme importance on Apollo 16's payload is the makings for the first astronomical observatory to be deployed on another planet. This is significant because it will permit observation and photographic work to be undertaken without the inhibitions found when looking through an atmosphere. Looking at Apollo 16 in a broader perspective, we see that this mission will form a link in a chain of scientific stations consisting of the various Apollo landing sites on the moon's surface. Certain instruments remain operative long after the astronauts return home, and listeners on earth will continue to draw data from them for many years to come. We are witnessing in Apollo 16 the continuing formation of the most monumental scientific observations ever conducted. Our thrusts outward lack the hesitancy of just a few years ago, as well as the doubt.

JOHN W. YOUNG

SPACECRAFT COMMANDER/ Born in San Francisco, California, September 24, 1930. Captain, USN. Degrees: BS in Aeronautical Engineering from Georgia Institute of Technology; Honorary PhD in Law from Western State University College; Honorary PhD in Applied Science from Florida Technological University. Flight experience in excess of 6300 hours. Prior NASA work: pilot, Gemini 3; backup pilot, Gemini 6; command pilot, Gemini 10; backup CM pilot, Apollo 10; prior to Apollo 16 he was backup Commander, Apollo 13.

T. KENNETH MATTINGLY, II

COMMAND MODULE PILOT/ Born, Chicago, Illinois, March 17, 1936. Commander, USN. Degree: BS in Aeronautical Engineering from Auburn University. 4700 hours flight time. Prior assignment to NASA was the Air Force Aerospace Research Pilot School where he was a student. Served aboard the carriers USS *Saratoga* and *Franklin D. Roosevelt,* he gained most of his flight experience flying both A1-H and A3-B aircraft. Scheduled as CM pilot for Apollo 13, he was medically grounded before launch with a case of measles.

CHARLES M. DUKE, JR.

LUNAR MODULE PILOT/ Born, Charlotte, North Carolina, October 3, 1935. Colonel, USAF. Degrees: BS in Naval Sciences from the U.S. Naval Academy; MS in Aeronautics and Astronautics from MIT. Has logged 3500 hours flying time. Graduate and former instructor at the Air Force Aerospace Research Pilot School, where he specialized in control systems and flight instruction in F-104, F-101, and T-33 jet aircraft. Prior NASA experience includes service in the support crew of Apollo 10, and backup LM pilot for Apollo 13.

APOLLO 16
YOUNG•MATTINGLY•DUKE

APOLLO 16 Preparation

Apollo 16 is yet another excellent example of how well scientists and engineers worked together to get the most out of lunar exploration. Although Apollo 16 will be the fifth time men have landed on the moon, this expedition was planned and trained for with as much preparation as previous expeditions. Something new is learned during the preparation and execution of each mission, and Apollo 16 benefitted from all the experience and knowledge already gained. John Young (**1**), in full moon gear, practices deploying the Lunar Roving Vehicle from its stored position in the LM at Kennedy Space Center. Young and Duke (**2**) collect rock samples along their simulated lunar traverse during a geology field trip in the Coso Hills near Ridgecrest, Calif. Mission Commander Young, kneeling, (**3**) checks out the surface magnetometer he will deploy on the moon's surface when he and Duke (in orange coveralls at right) land on the moon on April 20, 1972. Backup Mission Commander Fred W. Haise, Jr., watches at left. The LRV qualifi-

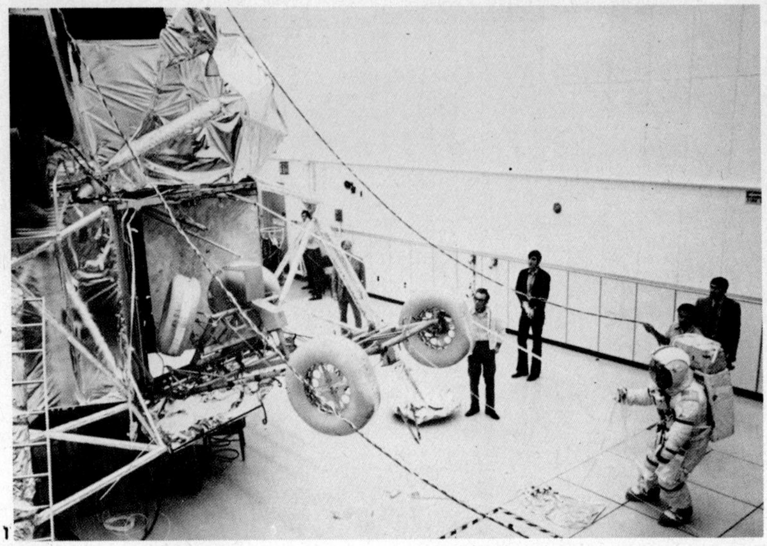

1

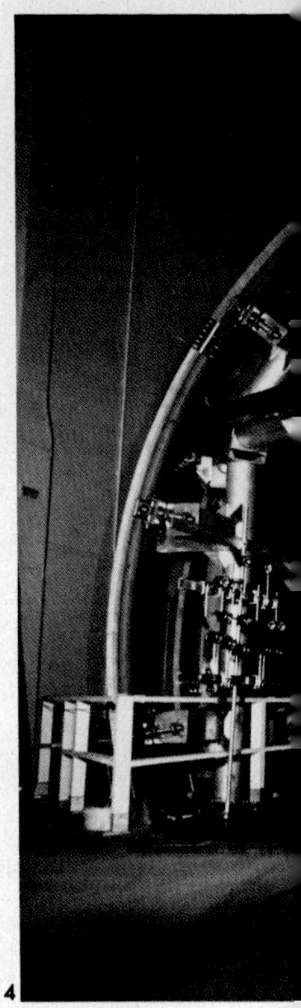

4

2

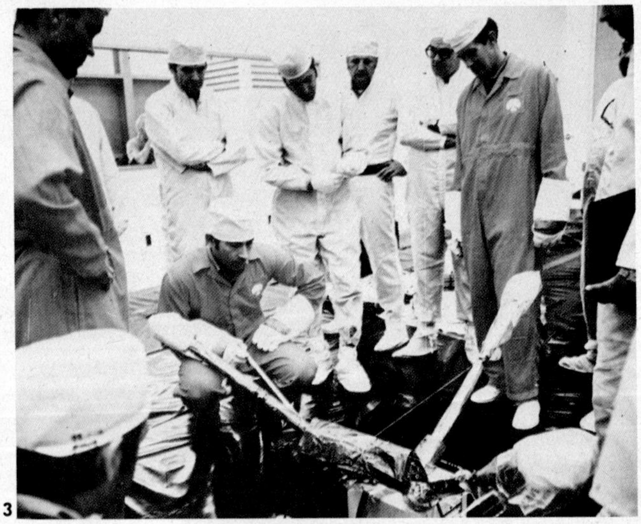

3

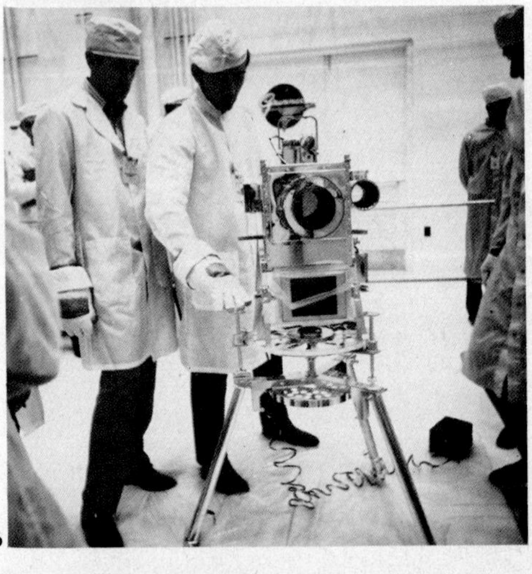

6

7

cation unit, (**4**) identical in almost every respect to those designed for actual use on the moon, underwent a series of magnetic property tests at Goddard Magnetic Test Facility, Greenbelt, Md., months prior to the moon landing. The LRV's magnetic properties were checked to determine the length of cable required for the portable magnetometer to be carried on the Apollo 16 mission; tests showed that the magnetic seals of the LRV and auxiliary equipment extended 46 feet from the LRV. The Apollo 16 flight crew in the CM Simulator (**5**) undergo a mission profile at Kennedy Space Center. Mattingly and Young are inside the CMS. Engineers at Kennedy Space Center (**6**) check out and operate a lunar surface telescope. The astronauts will set up and operate this telescope at Descartes to gather photographic, imagery and radiation data on the earth's upper atmosphere, aurorae and corona, and to map the location of interplanetary, interstellar and intergalactic hydrogen. LM Pilot Duke (**7**) removes his spacesuit gloves while talking to CM Pilot Mattingly following mission simulator exercises. Duke (**8**) undergoes a final suit check at the International Latex Corporation's plant, Dover, Delaware, where an adjustment was made to the waist joint of his 7 lb. EVA suit. The Apollo 16 space vehicle (**9**), was rolled out of the VAB upon its mobile launcher at 7 AM EST, 12/13/71. The 3.5-mile trip to Pad 39A took 6 hours, and the space vehicle underwent a series of tests before going into the critical flight readiness and CDDT phase.

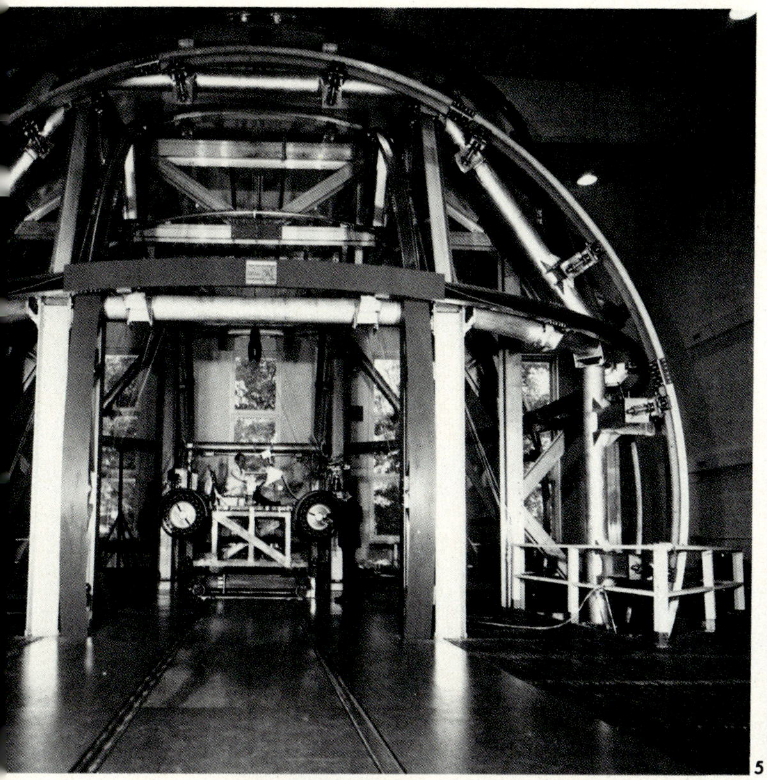

5

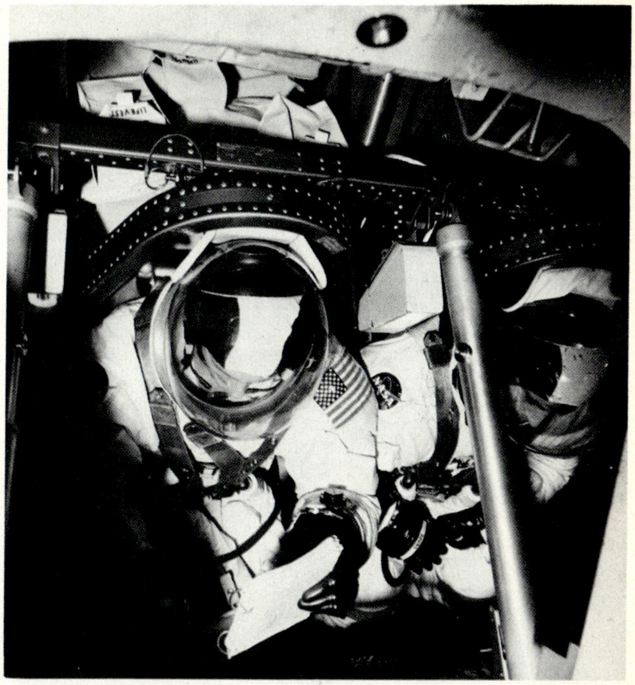

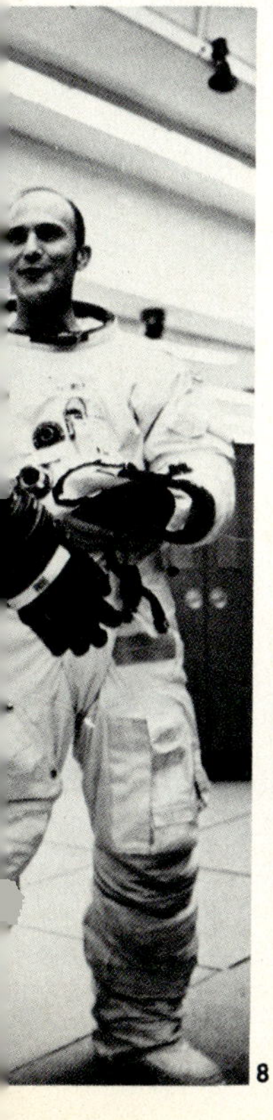

8

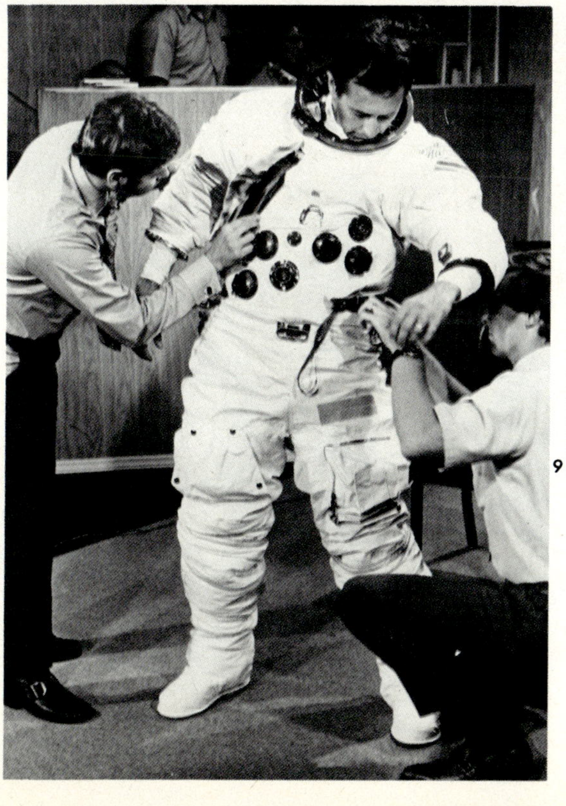

9

APOLLO 16
Launch

Extensive pre-launch training and operations have been completed, and the next step is the launching. This was to be the last daytime lift-off of an Apollo flight. The usual excited throng of spectators crowded the cape to watch the Apollo 16 lift-off on Sunday, April 16, 1972. (**1**) Young, Duke and Mattingly (left to right) enter the transfer van at the start of their brief ride to the launch pad and the Saturn V rocket, scheduled to take off at 12:54 AM EST. Approximately 2,000 accredited newsmen (**2**) witnessed and photographed the Apollo 16 lift-off from the press site, located 3½ miles from the launch pad. The tradition of giving nicknames to the Command and Lunar Modules continues: the astronauts named their LM *Orion*, after the constellation Orion, and their CM was named *Casper*, after the popular TV cartoon, "Casper The Ghost". Since photographs of spacecraft taken on the lunar surface and in lunar orbit appear to have a ghostly or other-world look about them, *Casper* seems an appropriate and humorous

1

2

3

choice of name for the CM. Vice-President Spiro T. Agnew, on the right (**3**), NASA Administrator Dr. James T. Fletcher, center, and Deputy NASA Administrator Dr. George M. Low are present for the Apollo 16 lift-off and will monitor communications from their viewing area within the Launch Control Center. Searchlights flood the Apollo 16 Saturn V space vehicle and Pad A at Launch Complex 39 (**4**) during the "wet" or fueled portion of the Countdown Demonstration Test. The week-long CDDT is conducted as a dress rehearsal for the actual countdown. During CDDT, all major spacecraft systems and ground support equipment are verified. The astronauts participated in the "dry" or unfueled portion of the CDDT, monitoring spacecraft systems in the same manner as they would on launch day. Nothing is left to chance, and the astronauts are thoroughly trained in all aspects of the launch, flight and exploration before all systems are declared "Go". The Apollo 16 Saturn V space vehicle streaks skyward on schedule at the start of NASA's eighth manned voyage to the moon (following page). To the myriad machinists, assemblers, inspectors and technicians in NASA's far-flung network of contractors, which at peak involved more than 20,000 firms employing more than 350,000 people, A/S #511 with Spacecraft #113 and LM #11 was on its way to the moon. And with it went a part of each and every one of them. An overwhelming sense of pride and accomplishment filled the heart of Americans for the sixth time in three years!

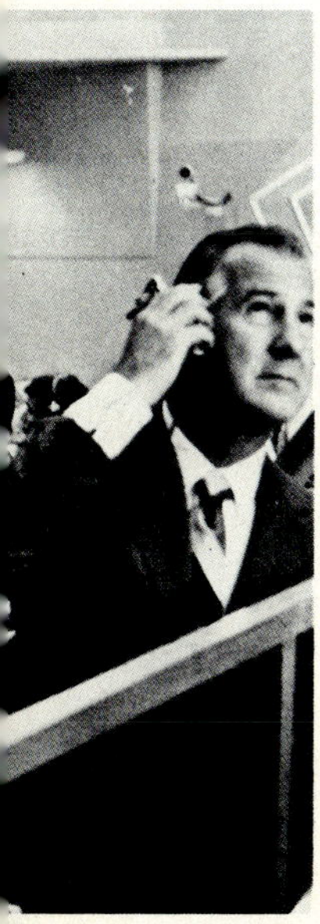

4

APOLLO 16
Launch

APOLLO 16
In-flight

After liftoff, the surrounding tension of which was heightened by a troublesome yaw rate gyro, trans-lunar insertion is achieved. To the annoyance of Mission Control and the astronauts, the last-minute ground problem marks a beginning for a series of in-flight problems. When *Orion* and *Casper* are established in a lunar trajectory, the crew reports that the high-temperature paint on the LM is blistering and subsequently flaking off. The paint is there for a reason: to protect the LM from the intense heat of the sun. It was determined that the Lunar Module could withstand the radiation without the coating, and operations proceed. The S-IVB third stage, now of no further use, is sent on a collision course with the moon to initiate seismic reaction. En route, its tracking system "dies," and Mission Control is no longer able to pinpoint its impact time and location. No mission into deep space is a routine matter. Apollo 16's destination (**1**) is captured in the stark clarity of space, looming larger through the windows of the Com-

2

1

3

mand Module. As the vehicle approaches lunar orbit, the lifeless landscape greets the crew (**2**). Visible in this oblique view is the Sea of Tranquillity, where, less than two years ago, the first men set foot on the moon. The crater with bright rays extending outward is Proclus, and near the left center edge is the crater Taruntius, located in the north-east portion of the Sea of Fertility. On the horizon at right is a portion of the Sea of Serenity. A closeup view (**3**) of King crater area shows features that have been changeless for ages. More trouble. For six hours, *Orion* and *Casper* orbit together after undocking, and the actual landing is almost called off. Mattingly, Mission Control, and NASA contractors on earth attempt to unravel a vibration in the CSM's engine when the secondary steering system is activated. Finally (**4**) the LM swings into descent mode with Mission Control's OK. But the time spent determining the problem won't hurt the CSM's capabilities has shortened the Apollo 16 excursion by one full day. A new flight plan is approved; several photographic missions involving attitude alterations are ditched. Commander Young says that the mission is a "cliff hanger" until the landing go-ahead is given by Houston. Three additional orbits of the moon have passed; its riddled face (**5**) reveals a mound feature impassively. A 250mm lens takes in the southwestern rim of King crater during orbit (**6**); the distance from the rim to center peak is about 30 kilometers. The hook-shaped object is a piece of debris on the camera lens.

5

6

APOLLO 16
Exploration

"Give me a salute!" barks Duke, and Young responds with a show of one-sixth gravity enthusiasm, leaping straight up. The television camera records this scene (**1**) for millions of viewers on earth. At the same instant (**2**) Charles Duke's camera records the amusing event at Descartes Base, close observation shows Young's boots clearly off the surface. Touching down on the Cayley Plains in the Descartes Region makes Apollo 16 the most southerly site visited on the moon, as well as the highest—about 2400 meters (8000 feet) higher than Apollo 11's base, some 240 km (150 miles)NE. In dazzling light, John Young (**3**) works with the lunar surface drill at an ALSEP deployment. In the foreground is the Lunar Surface Magnetometer. For the first time in history astronomical observations are taken from another world (**4**). The S-201 Lunar Surface Ultraviolet Camera/Spectrograph is placed in the shadow of the LM. Built by the Naval Research Laboratory, the 3-inch electronographic camera will reveal and photograph

1

invisible ultraviolet light from the earth and its geosphere, as well as from other celestial bodies. Observation such as this is impossible on earth, due to the atmospheric blanket. A spectacular example of celestial U/V photography (**5**); color enhancement of the earth's upper atmosphere reveals our planet in a "new light". The bright belts extending outward are far-ultraviolet emissions at a wavelength of 1216 Angstrom units. The apparently knobby vertical projection represents several ultraviolet emissions from the sunlit atmosphere, each bulge being produced by one type of gas. Spectral dispersion is approximately 10 Angstrom units per millimeter on this plate. Researchers hope to not only locate new hydrogen clouds in the universe, but also to gain deeper insight into the atmospheric composition. When Young and Duke set up a geophysical research station 300 feet from their landing site, they have brought the total number of automated instruments on the lunar surface to 20. The four stations form a triangular network in which each side of the triangle measures about 1000 km (600 miles). Being rathe close, Apollo 12 and 14 sites are taken tc be one point. This system fulfills a centuries old hope of scientists: making long-range measurements and determinations from another world. Apollo 11's only surviving in strument is a laser retro reflector. Charles Duke not only gathered souvenirs from the moon, but left one of his own (**6**) in the form of a family portrait encased in plastic. I will rest undisturbed there for centuries

4

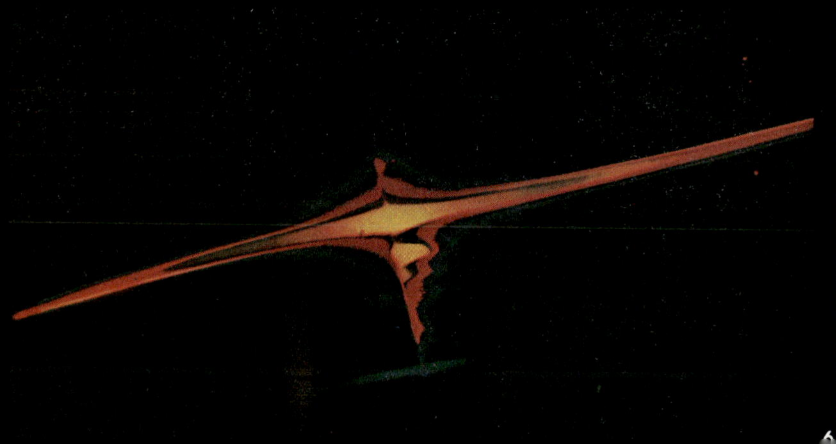

6

APOLLO 16
Exploration

Lunar module pilot Charles M. Duke Jr., doesn't waste a chance to show off "his ship", *Orion* (**1**), next to the lunar rover. Astronaut John W. Young appears directly behind the LRV. Another EVA-1 orientation photo (**2**), features the lunar rover at Station 4, near Stone Mountain, during the mission's second extravehicular activity (EVA-2). Rocks in the area show angularity and Palmetto Crater can be seen just over the top of the rake on the rear of the rover—about 1½ inches above the right rear wheel. Light rays from South Ray Crater can be seen in the background, to Duke's left. But a sample of the terrain near the Descartes landing site is unmistakable, as Duke ambles along, lunar style, at the rim of Plum Crater. The eastward view of this crater (**3**) shows part of the impacted dimensions, nearly 40 yards wide by 30 feet deep; the rover can be seen parked on the other edge. Mission Commander Young changes roles from photographer to prospector, as he uses the surface rake (**4**). A gnomon, a measuring device,

2

3

which locates the local vertical and is used in later photographic evaluations, is at his right foot. Duke, meanwhile, is kept busy, inspecting the surface of a large boulder at North Ray Crater during EVA-3. Note his chest-mounted 70mm Hasselblad camera (**5**). All is not a slow operation, however, as the "cross-country" capabilities of the LRV are making test drivers of the astronauts. Imagine a slalom course on the beach, when you don't have to worry about other cars, since the only other car on the planet is more than 600 miles away, and parked! The distance between Apollo 16's landing site at Descartes and the Hadley-Apennine site of Apollo 15 is about 1,000 kilometers or 600 miles. LRV-2, however made up for a two-car race, by flexing its battery-powered muscles in style (**6**). Duke's 16mm Maurer camera records a trail of flying moon dust as Young throws the rover about in what was called the lunar Grand Prix (**7**). The object of this test was to check wheel adhesion and other driving characteristics; note that the right front wheel's off the surface. Young is driving at speeds up to 11 kilometers (seven miles) per hour over small craters, causing the front wheels to bounce off the surface. Meanwhile, in lunar orbit, Command Pilot Thomas Mattingly awaits docking. The following sequence (**8**) shows *Orion's* ascent stage as it pirouettes for Mattingly's inspection prior to docking. Note skin deformation of the ascent stage's thin panels, not an unusual occurrence after lunar liftoff. Docking took place at 10:47 PM.

5

7

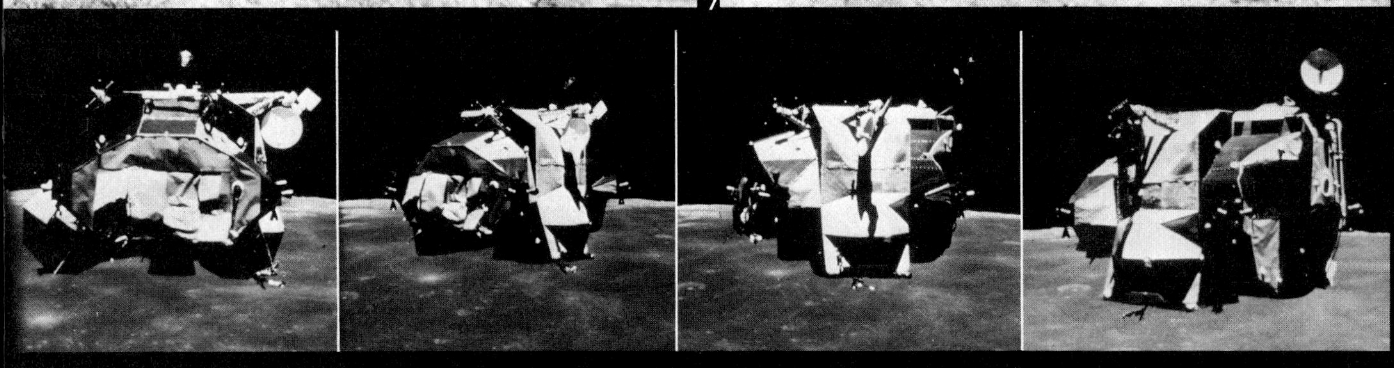

APOLLO 16 Splashdown

Another extravehicular activity (**1**) is performed during the Apollo 16 trans-earth coast; Mattingly climbs out of the CM to inspect the SIM bay of the Service Module and retrieve its film cassettes. He is wearing Young's helmet, because the lunar helmet's extravehicular assembly helped protect his eyes from the bright sun. Duke, on the left, can be seen leaning from the spacecraft hatch to assist Mattingly. The flight back from the Moon is uneventful, and Apollo 16 ends safely with a textbook splashdown in the Pacific at 2:45 PM EST, April 27, 1972, at a point about 1,500 nautical miles south of Hawaii. The three-orange-and-white parachutes (**2**) blossom against the bright sky, and slowly lower the Apollo spacecraft a few miles from the carrier *USS Ticonderoga.* Mattingly (**3**) is hoisted aboard a helicopter during recovery operations after splashdown. Young and Duke are plucked out of the rubber raft beside the CM and into the helicopter in the same manner. It is standard procedure for Navy pararescuemen to jump

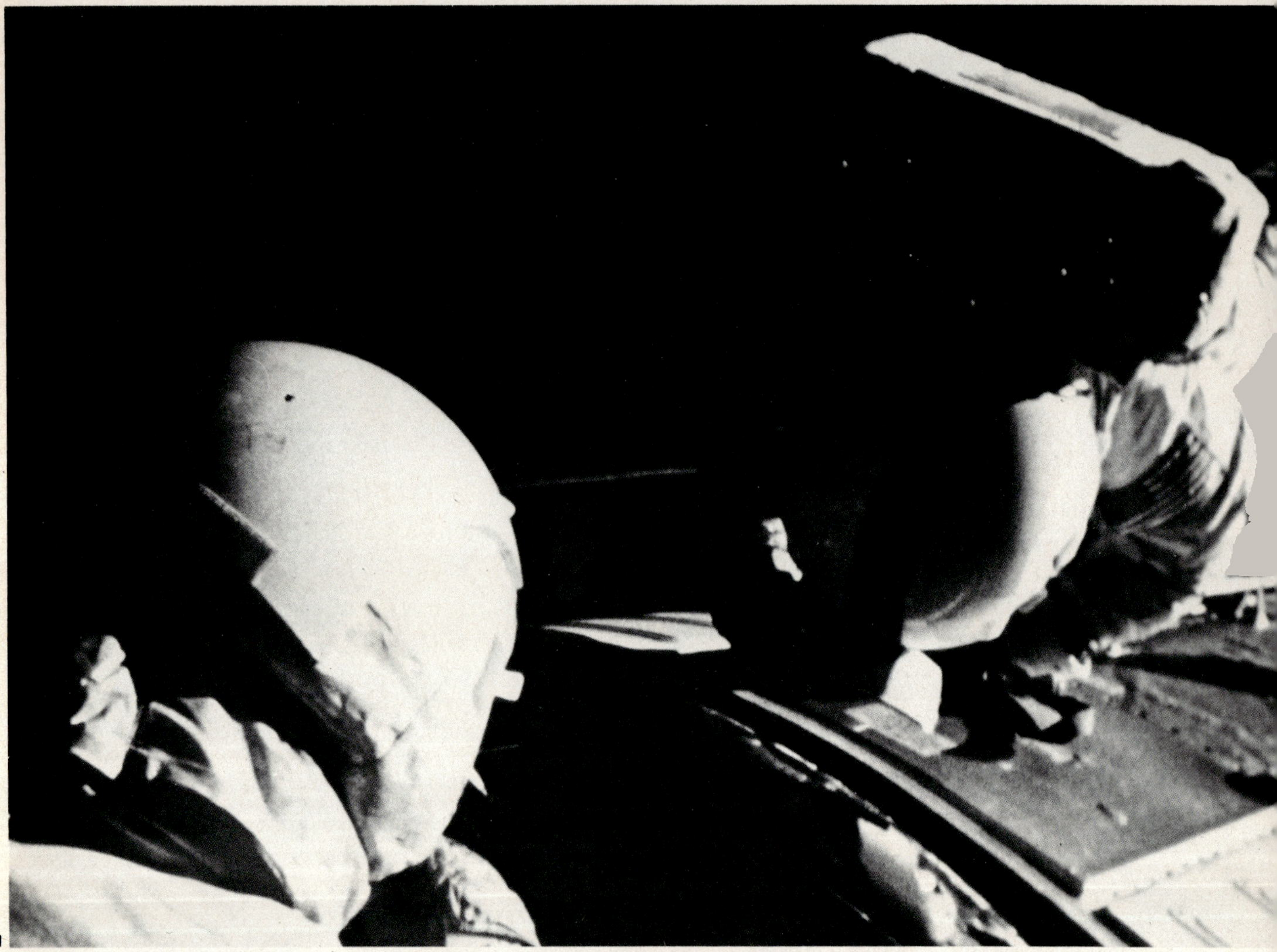

1

2

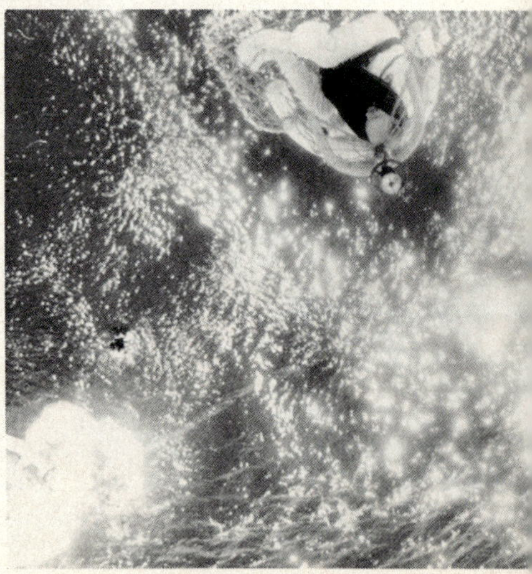

3

into the ocean from a helicopter at the start of all recovery operations. They attach the CM flotation collar, bring the life raft up close, and assist the astronauts out. The recovery operations for Apollo 16 are carried out as precisely as all previous Apollo missions. Two of the men (**4**) can be seen just as they make their jump. Performing with their usual speed and efficiency, the pararescuemen and helicopter pilots have the three astronauts safe on the deck of the *Ticonderoga* within minutes after their return from the moon. Each Apollo mission extends the engineering and scientific knowledge based on the combined experience of all preceeding missions. The Apollo 16 astronauts lived on the moon longer than any other astronauts, spent more time exploring the lunar surface, and brought back more geological samples—213 pounds—than the crews of any of the previous flights. This specimen (**5**) is one of the larger samples, a medium gray anorthosite-type rock that measures 18x16x7 centimeters and weighs 1185 grams (2.6 pounds). One of its notable features is an almost complete lack of cavities, or vugs, most unusual in a non-sedimentary rock. Geologists term this rock "crudely slabby, subangular to angular", and speculate that the gases present during its formation were trapped in its intergranular spaces. Summing up this mission, John Young quoted the great man himself, Rene Descartes: "There is nothing so far removed as to be beyond our reach or so hidden that we cannot discover it." A fitting commentary.

4

5

Rightly called "the voice of the moon," the Apollo Lunar Scientific Experiments Package (ALSEP) is an example of scientific and manufacturing sophistication. ALSEP is investigating the evolution of the moon and the solar system through instruments such as seismometers, magnetometers, temperature probes, and ion detectors. On the threshold of Apollo 17's mission, this ALSEP network of four lunar observatories has accepted and executed more than 38,000 commands from earth, and has returned at least 23 billion data measurements. Now, every time a grapefruit-size meteorite hits the moon, an instant recording appears on earth. Each ALSEP is good for more than one year, with Apollo 17's ALSEP-E designed for a two-year life. Apollo 11's instruments differed from subsequent packages mainly in the power source. The Early Apollo Scientific Experiments Package (EASEP) was used only once (Apollo 11), and utilized solar panels in place of the later little nuclear reactor called the Radioisotope Ther-

ALSEP CHRONOLOGY

Instruments left on the moon are still relaying data to earth

ILLUSTRATIONS BY KEN HODGES

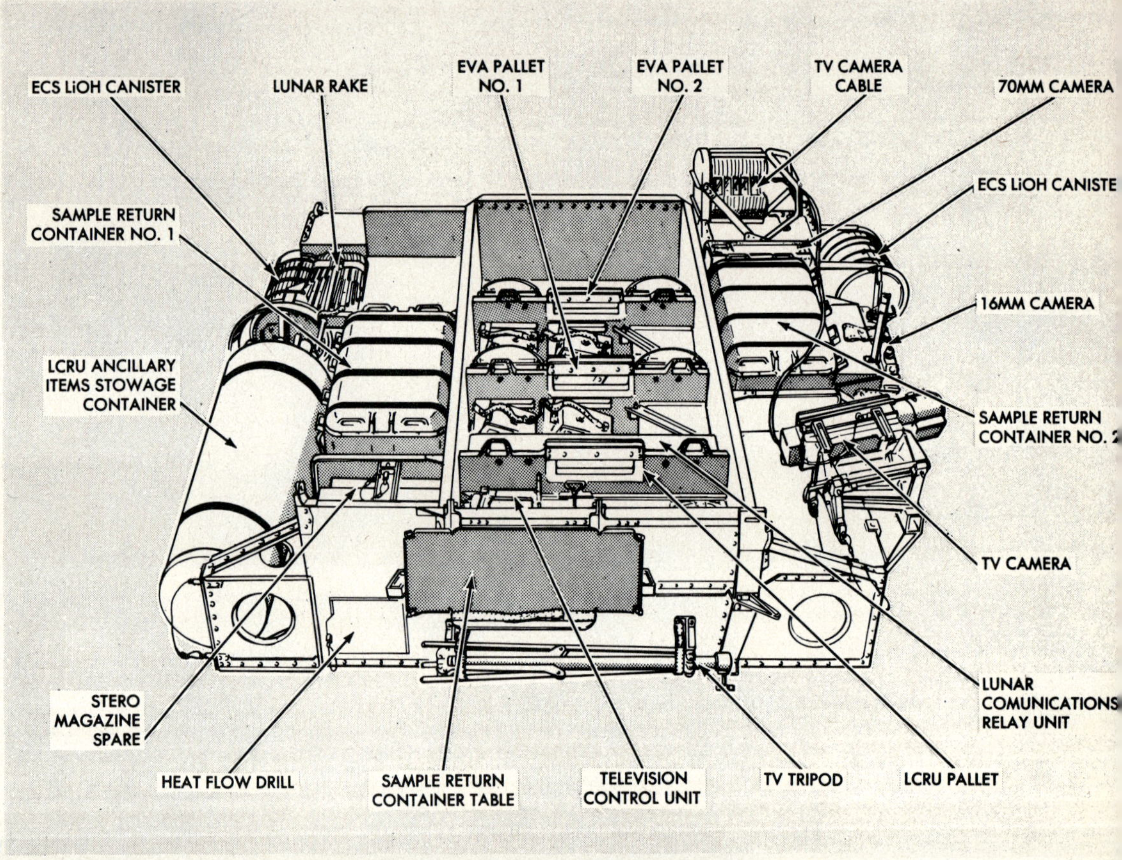

1

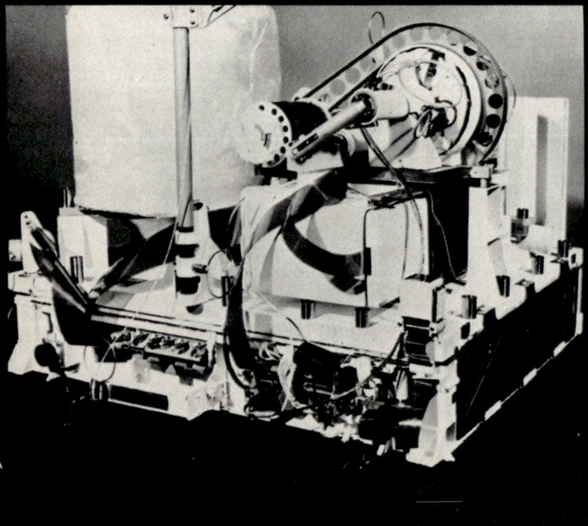

A

B

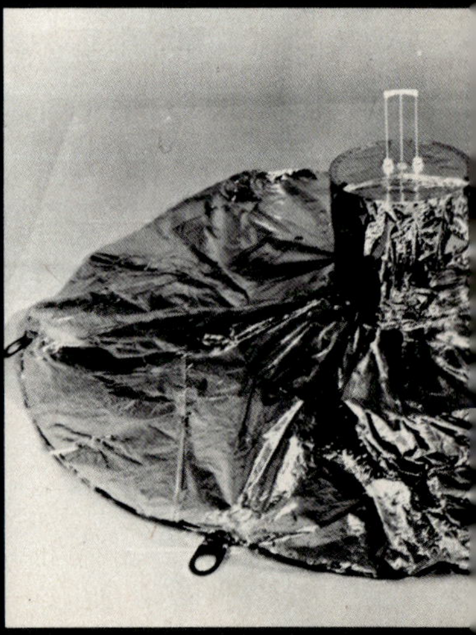

C

moelectric Generator—also known as the Space Nuclear Auxiliary Power (SNAP-27). EASEP was designed to operate for only one lunar day (about 13 earth days), but continued to work for four and one-half months—quite an achievement, considering that lunar daytime temperatures hover near 250 degrees Fahrenheit and drop at night to minus 300 degrees. The newer package, the ALSEP, weighs 285 pounds, and is transported to the moon in two tightly-packed pallets, which occupy no more than 15 cubic feet. A third pallet, the Modularized Equipment Stowage Assembly (MESA), is stowed in the Lunar Module descent stage, like all ALSEP equipment. MESA (**1**) contains some auxiliary ALSEP tools, along with cameras and specimen storage containers. Structural and thermal components of ALSEP are stowed in subpackage 1, shown here (**A**) with the Thumper/Geophone Assembly cover off. Next to it is a large covered drum, the Passive Seismic Experiment. Both of these experiments measure seismic disturbances; the latter detects natural moonquakes, while the "Thumper" produces artificial ones. The EASEP package (**B**) used only the passive seismometer, which has an unmistakable appearance (**C**); as does the Lunar Surface Magnetometer, which is designed to record the moon's slight magnetic field (**D**) The Laser-Ranging Retroreflector (**E**) uses its 1 to 300 high-purity quartz mirrors to bounce laser beams back to earth in order to gauge earth-moon distances to an accuracy of 6 inches.

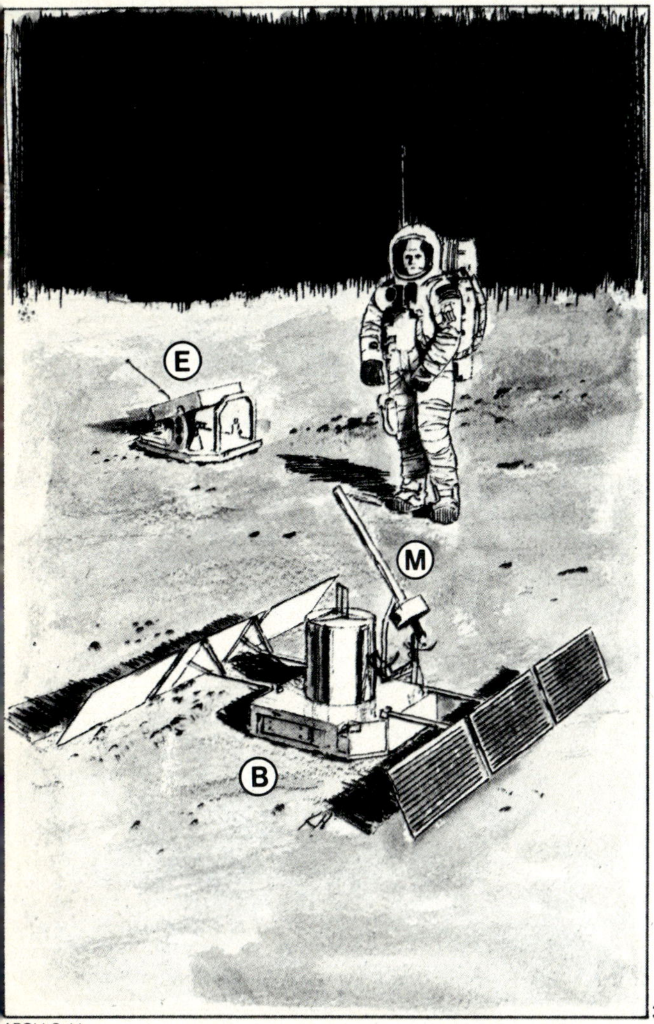

APOLLO 11

3

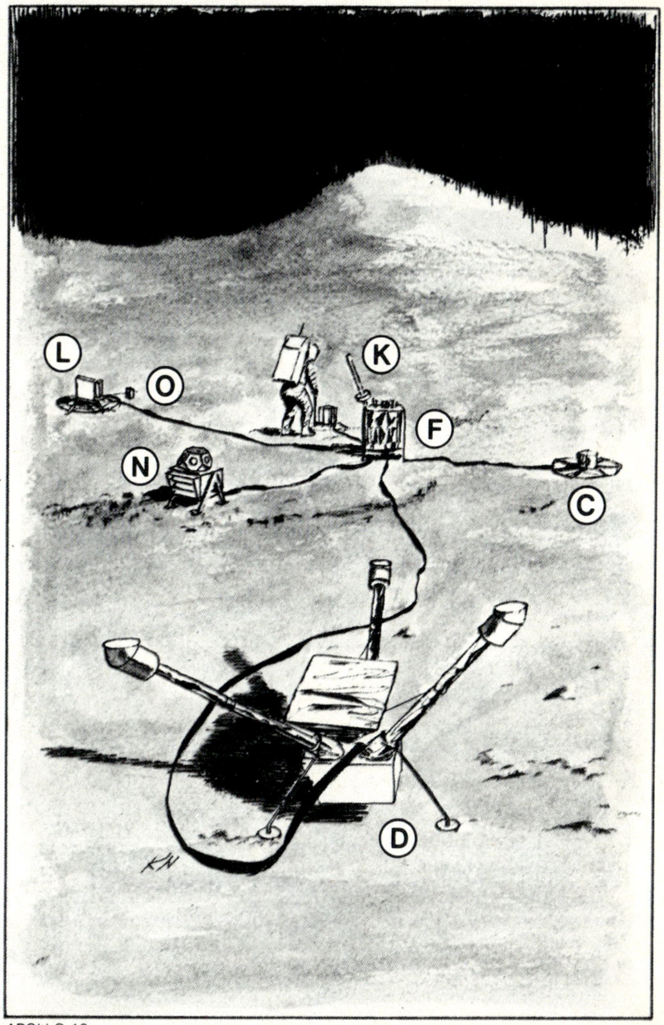

APOLLO 12

D

E

COURTESY BENDIX CORP.

ALSEP

Serving as the clearing house for all lunar data is the ALSEP Central Station (**F**). It collects and formats data from the experiments, monitors performance of all its systems, and transmits these to earth as pulse-code modulation on an S-band telemetry signal. The S-band antenna must be aligned properly, once in position (**M**). The heart of the Central Station is a shielded capsule of plutonium-238, which provides 63 to 74 watts of power for more than one year. On the moon, the fuel capsule is removed from its storage place and then inserted into the SNAP-27 generator (**K**). Stowed in subpackage 2, the power generator's eight fins give it a paddlewheel appearance; these help to stabilize its temperature. First used with Apollo 15, the Heat Flow Experiment (**G**) takes temperature and conductivity measurements in the moon's subsurface. Apollo 15's package used the same instruments as Apollo 12, except for the additional Heat Flow Experiment and a larger 300-mirror Laser Retroreflector. All other ALSEPs use

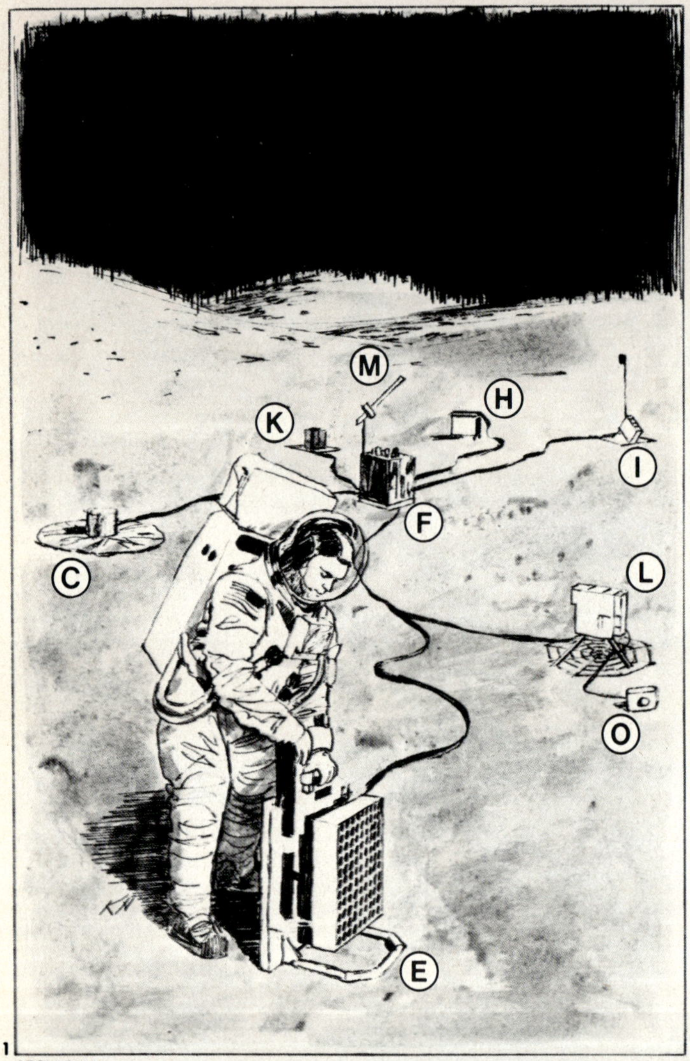

1 APOLLO 14

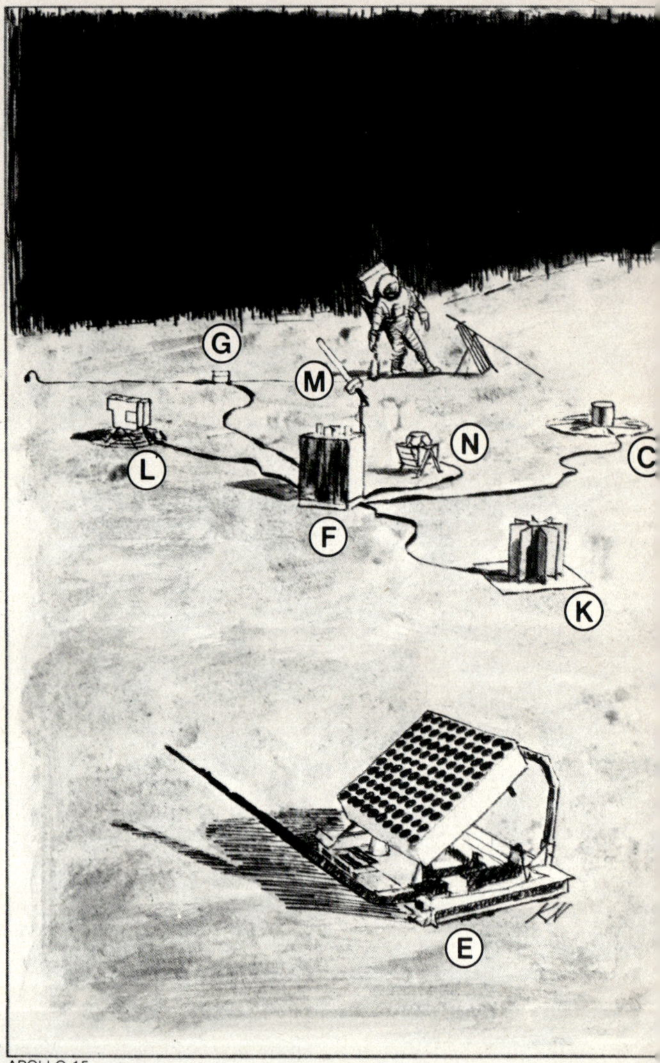

2 APOLLO 15

F

COURTESY BENDIX CORP.

G

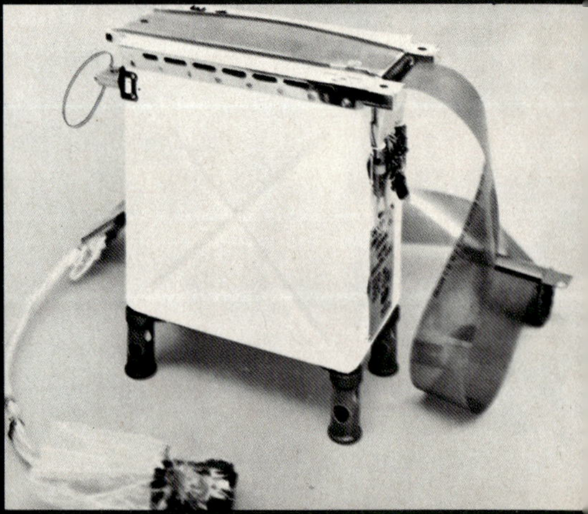

H

different instrument combinations, including several repeated experiments. The Charged Particle Lunar Environment Experiment (**H**) monitors the flux of electrons and protons that bombard the moon's surface. Similarly, the Solar Wind Spectrometer (**N**) also measures bombarding particles, but includes velocity, direction, and temperature readings. The Active Seismic Experiment (**I**) was used for the first time on Apollo 14's ALSEP-C. The ASE is in reality two different experiments. The first part uses a small rod resembling a bar-bell—the Thumper—which fires 21 sound charges into the lunar soil, and the effects are measured through geophones, or listening devices, on the end of the rod. The other part reacts upon commands from earth to fire four rocket-propelled grenades to ranges of 500, 1000, 3000, and 5000 feet; when they explode, the shock waves are picked up by the passive seismometer. The Lunar Surface Gravimeter Experiment (**J**) can detect tiny changes in the moon's gravitational waves. The Suprathermal Ion Detector (**L**) registers the creation rate of ions (electrically charged particles) in the thin lunar atmosphere, and detects ions from space. Connected to it is the smaller Cold Cathode Ion Gauge (**O**), which monitors the lunar atmosphere density. The Lunar Ejecta Meteoroid Experiment (**P**) will determine speed, direction, mass, and density of cosmic dust particles. And the Lunar Mass Spectrometer (**Q**) will identify constituents of the lunar atmosphere. ALSEP, indeed, is a masterpiece of science.

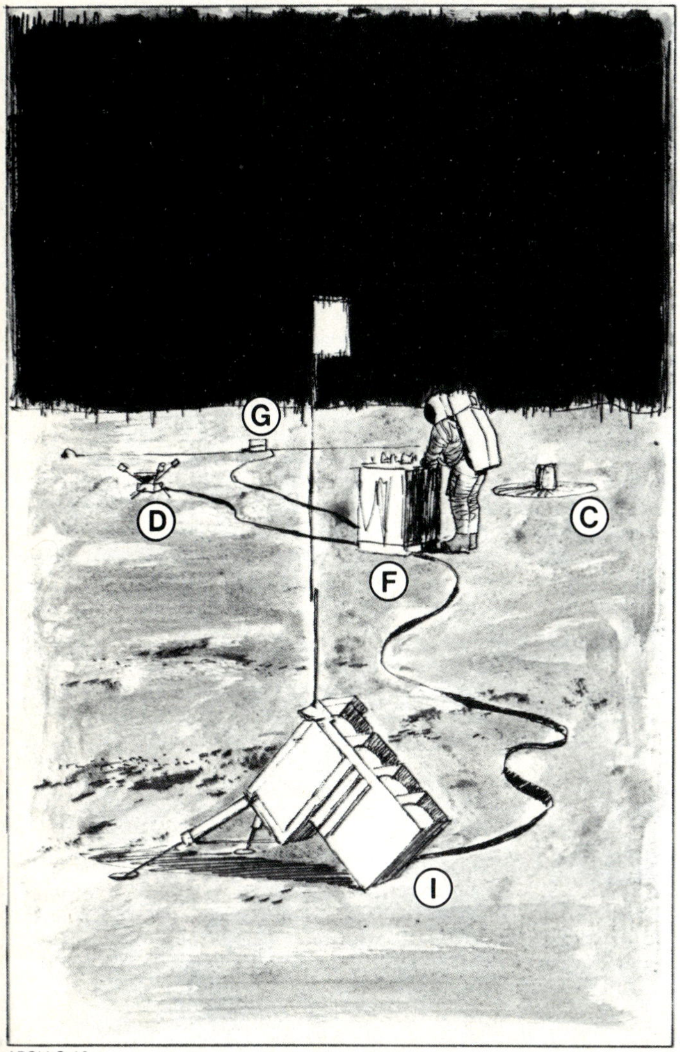

APOLLO 16

4

APOLLO 17

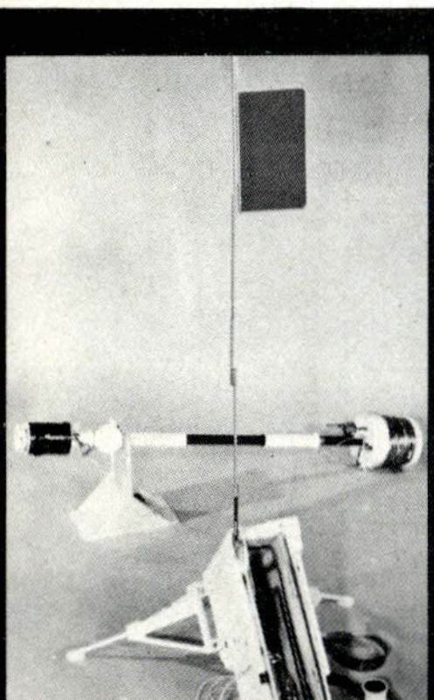
J

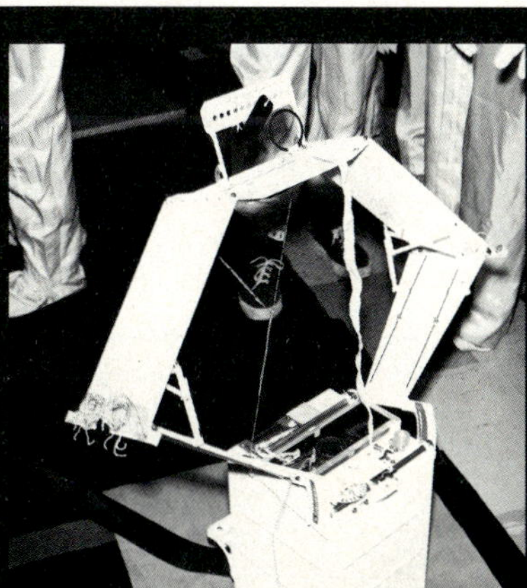
K

COURTESY BENDIX CORP.

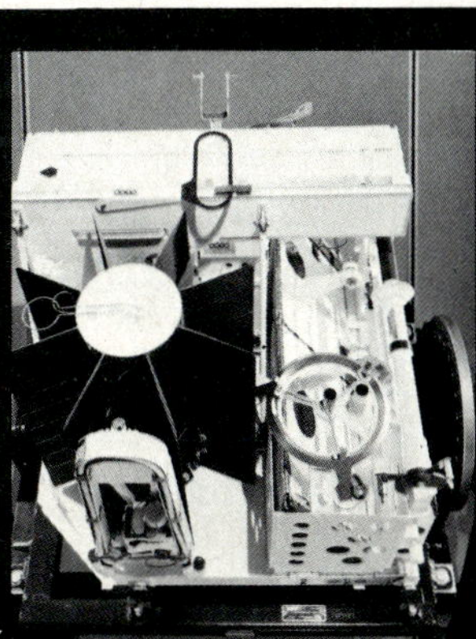

COURTESY BENDIX CORP.

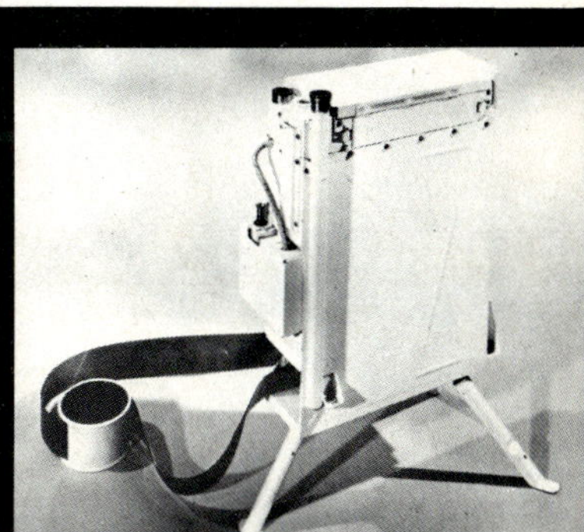
L

APOLLO 17

Apollo 17's emblem, designed with assistance from the crewmen, is the image of Apollo, Greek god of the sun; the visage is that of Apollo Belvedere, a sculpture in the Vatican City. Behind the head of Apollo is the outline of a modern American eagle, set in a background of deep blue, symbolizing space. The red bars of the eagle's wing symbolize the bars in our flag; the stars recall the three astronauts. Suspended in space are the planet Saturn and a galaxy of stars, as well as the moon. The eagle's wing partly bisects the moon, to indicate that this is the celestial body we have conquered. But the forward thrust of the eagle, the gaze of Apollo to the right toward Saturn and the galaxy, suggest that man's goals in space will someday include the planets and the stars. The colors, red, white and blue, are added to gold symbolizing the golden age of space exploration that includes the flight of Apollo 17. To many earthbound observers, Apollo 17 has an air of finality about it. True, it is the last manned exploration of lunar soil in the foreseeable future. But to consider it an end would be to detract from its significance. The data and wisdom that we will have gained from Apollo 17 and its predecessors will require years to comprehend.

EUGENE A. CERNAN

SPACECRAFT COMMANDER/ Born Chicago, Illinois, March 14, 1934. Captain, USN. Degrees: BS in Electrical Engineering from Purdue; MS in Aeronautical Engineering from the USN Postgraduate School; Honorary PhD in Laws from Western State; Honorary PhD in Engineering from Purdue. More than 4300 hours flight time. Prior NASA experience: pilot on Gemini 9; backup pilot for Gemini 12; backup Lunar Module pilot for Apollo 7; Lunar Module pilot for Apollo 10 lunar orbital flight; backup Commander, Apollo 14 Flight.

RONALD E. EVANS

COMMAND MODULE PILOT/ Born November 10, 1933, St. Francis, Kansas. Commander, USN. Degrees: BS, Electrical engineering, U. of Kansas; MS, Aeronautical engineering, USN Postgraduate School. Earned commission via Naval ROTC, U. of Kansas. Total flight time in service is 4000 hours. Combat flight instructor in F-8 aircraft from January, 1961 to June, 1962; flew seven months combat duty in Viet Nam. Prior NASA experience as support crewman on Apollo 7 and Apollo 11; backup CM pilot for the Apollo 14 mission.

HARRISON H. SCHMITT

LUNAR MODULE PILOT/ Born July 3, 1935, Santa Rita, New Mexico. Degrees: BS, Science, Cal Tech; studied at University of Oslo, Norway; PhD, Geology, Harvard University. Served with USGS in the Astrogeology department instructing NASA astronauts. Logged in excess of 1500 hours flight time after entering astronaut training. Prior NASA experience: instruction of other astronauts in science, navigation and field studies; integration of scientific activities into Apollo missions; research of lunar rock; backup LM pilot on Apollo 15.

APOLLO ★ XVII
CERNAN ★ EVANS ★ SCHMITT

APOLLO 17
Pre-launch

Among the specialized training given the astronauts is a briefing on lunar geology. Ronald Evans (**1**) surveys lunar maps during a session with Farouk El-Baz, Supervisor, Lunar Science Planning, Bellcom, Inc. To the instructor's left is astronaut Robert Overmeyer, a member of the Astronaut Support Team for Apollo 17. Despite the inclusion of Harrison Schmitt in the mission, who holds a doctorate in geology, a practical knowledge in the field is vital to all astronauts. The stark landscape near Boulder City, Nevada, provides the training ground for Apollo 17 astronauts Schmitt and Cernan (**2**). They are using the training version of LRV, equipped with off-road tires. Informally dressed, they nonetheless carry PLSS systems. Schmitt, being a geologist, is "at home" studying a terrain map (**3**), while Cernan looks on. Both are carrying Hasselblad lunar cameras, behind which can be seen specimen bags. An idea of the size of the PLSS units can be obtained. Radio contact is maintained with the training group at

1

2

3

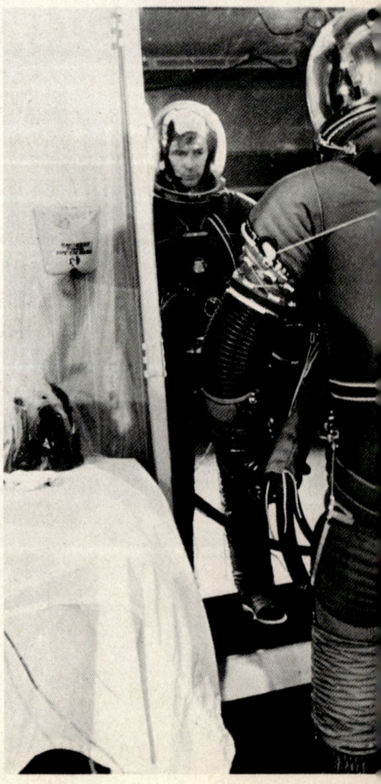

4

all times. At ILC Industries plant in Dover, Delaware, (**4**) Gene Cernan stands in front of a "tailor's" mirror during a final fitting session of his flight suit. In spite of its bulky look when pressurized, the accordion-like sections around arms and legs allow freedom of motion in the CM. Lunar Module Pilot Schmitt (**5**) verifies the carefully stowed material that will be carried aboard the LM to the moon. At Kennedy Space Center (**6**) the astronauts undergo testing of their EVA suits in the altitude test chamber. One technician checks the boot of Schmitt's suit, as Cernan hands him a slip of paper. Another technician records the test sequence on 8mm film. Carried out at Complex 39, in July, 1972, the altitude chamber sequence is among the final phases of training and testing for Apollo 17. Harrison Schmitt (**7**) cavorts around at ILC Industries plant during flight fit checkout and integration tests of the Apollo 7-L spacesuit. While Schmitt amuses the crew with his dexterity in the bulky EVA suit, an assistant holds the oxygen tubes clear. Schmitt is the first astronaut to hold a doctorate in a field relevant to lunar science. Since Apollo 17 is the final manned mission to the moon, NASA has not formed another three-man astronaut team. Therefore, the backup crew for Apollo 17 consists of space veterans John Young, commander, who was CM pilot for Apollo 10 and commander of Apollo 16; CM pilot Stu Roosa, who filled that role for Apollo 14 and was also backup for Apollo 16; and LM pilot Charles Duke, also of the Apollo 16 crew.

6

5

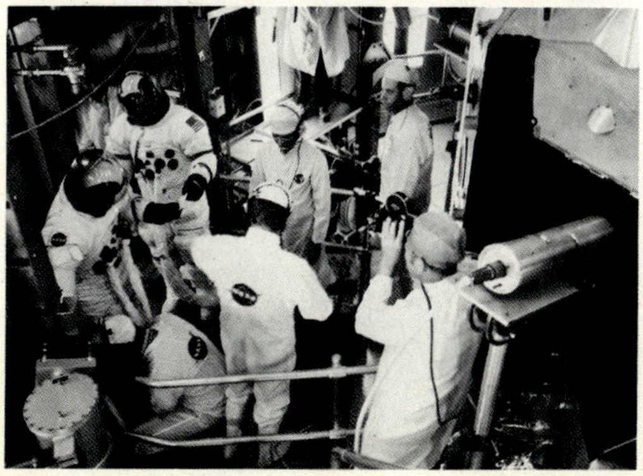

7

APOLLO 17
Pre-Launch

Astronaut Eugene Cernan (**1**) makes a point during a press conference at NASA's Manned Spacecraft Center in Houston. He is flanked by Harrison Schmitt, left, and Ronald Evans. Later, the crew concluded emergency egress practice procedures at the Cape. The crew (**2**) is standing in one of the walkways of the Launch Umbilical Tower at Complex 39A. Should there be an emergency or failure during or before launch, several "ditching" procedures are available to the astronauts. A 250-ton gantry crane (**3**) gently lifts the mighty first stage Saturn 5A (#512) off its trailers in the Vehicle Assembly Building at the Cape. Note that the gimballed F-1 engines are locked to the hull for transport. In the Vehicle Assembly Building (**4**), the Lunar Module, serial #12 and the Command and Service Module, serial #114, undergo a docking test under the watchful eyes of Cape technicians. The docking probe, visible in this photo, ensures alignment of the two vehicles and fits into the LM. Above and to the left is the docking

1

2

4

probe T-shaped docking target, which is utilized by the Command Module pilot to optically align with the LM's ascent stage during docking. The blue coating on *America* is to protect it from damage during test. A group of technicians, along with the crew, carry out mission simulation check of the Lunar Rover (**5**) prior to final installation in the LM. The latter looks down imposingly on the undertakings. Practically all the LM's protective gold foil is in place. The Lunar Module, landing legs retracted, is prepared for attachment to the Command and Service Module (**6**). *Challenger* is raised by a crane into position. Designed for deep space function only, streamlining is provided by the Spacecraft Lunar Module Adaptor (SLA). The portion of the SLA surrounding the LM is jettisoned after the spacecraft is on its way to the moon, prior to the initial docking with the CSM. About 4½ hours before launch time, the prime crew (**7**) enjoy a relaxed dinner in the crew quarters at the Cape. Shortly, they will suit up and prepare to enter the *America* for the final lunar mission. "Hurry up, Ron," Gene Cernan seems to say (**8**) to Ron Evans, as the latter says goodbye to his wife. All three crewmen are in their flight suits, and carry portable air conditioning units for in-suit temperature modulation until in the CM. A van takes them from the Manned Spacecraft Operations Building to Complex 39A. The crew, once inside the Command and Service Module, will not leave the cramped confines of the spacecraft until on the lunar surface.

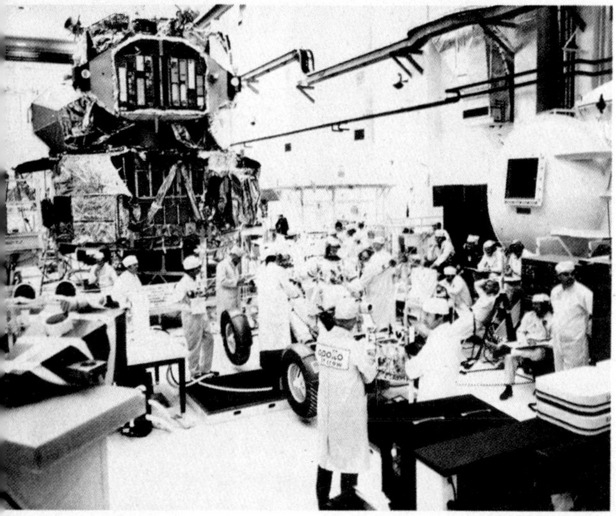

6

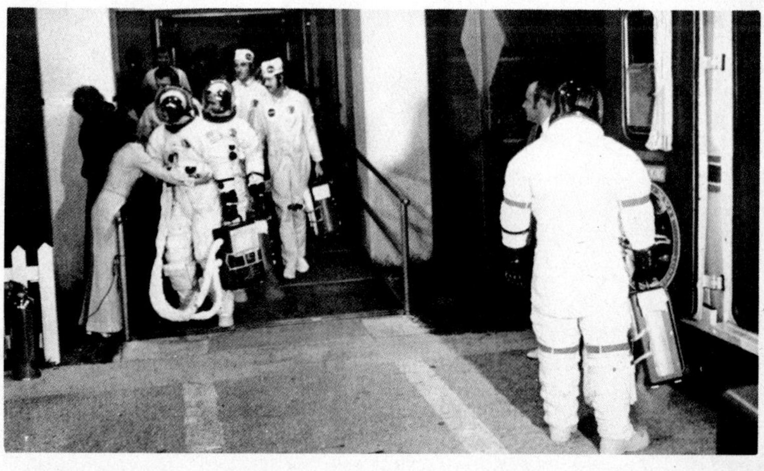

8

APOLLO 17 Launch

On August 28, 1972, doors of the Vehicle Assembly Building are opened, and the Apollo 17 launch vehicle (**1**) is slowly rolled out to Complex 39A on its transporter. Serialized Spacecraft #114, Lunar Module #12, Saturn #512 by NASA, its massive fuel tanks are still dry. The VAB, covering eight acres, is the world's largest building in volume. Containing 129,482,000 cubic feet of space, it is composed of a low-bay and high-bay section. The low-bay area contains eight stage assembly facilities and checkout cells. It is in the 525-foot-high main bay where the Apollo launch vehicle is assembled. Equally impressive is the 6,000,000-pound Transporter, two of which are in service at the Cape. They are the largest tracked vehicles in the world. They are propelled by 16 traction motors which receive power from four 1000Kw generators. Each transporter has four diesel engines, two of 2750 bhp for the 1000Kw generators, and two 1065 bhp engines which rotate a pair of 750Kw generators for the jacking, light

1

2

plants, steering and ventilation systems. The transporter slides under the portable Launch pad with attached Umbilical Tower and transports them, along with the Saturn to Complex 39A. Moving at a speed of one mile per hour, it is a lengthy excursion to 39A, 3.2 miles away. The silvery piping visible on the Umbilical structure is the fuel, water and environmental control system risers. They are coupled to ground connections at the pad. The launch vehicle, portable pad and Umbilical Tower weight in excess of 11,000,000 pounds. Once at the launch site (**2**), the spacecraft's systems are checked, followed by a space vehicle cutoff and malfunction test. Radio frequency compatability is set, followed by a flight-readiness test. Finally, a "dress rehersal" of the countdown is carried out, with the crew stationed in the CM. Only actual ignition is not carried out. Then the launch vehicle is recycled to pre-count status, and preparations for fuelling are made. The fuel is loaded under strict safety rules on the day prior to launch, beginning at around T-7 hours, requiring about 4½ hours. After fuelling, at about T-90 minutes, the crew boards the spacecraft. It is evening, and the launch pad is bathed it the light of 72 20Kw Xenon searchlights. In the Launch Control Center Firing Room #1 (**3**), every phase of the final countdown is monitored. During ignition and liftoff, 40,000 gallons per minute of water drench the pad, protecting it and the tower from near-total destruction. Despite this deluge, 39A always requires crews to do some "touch up" work.

3

APOLLO 17
Launch

APOLLO 17
Exploration

The mission's early stages go smoothly. Parking orbit is achieved, and three hours later the Saturn S-IVB begins its second burn, building up to escape velocity. During the six minutes of firing, a velocity of 10,000 feet per second is added to the spacecraft's orbital velocity, bringing the total speed to 34,250 feet per second. For Evans and Schmitt this is their first trip into space. Both men are elated at the view of Africa from parking orbit. Successful entry into lunar orbit comes at 2:48 PM EST on December 10.

The next day astronauts Jack Schmitt and Gene Cernan enter the LM and began their decent to the lunar surface. Touchdown at Taurus-Littrow comes at 1:55 PM on December 11. "The next thing it says is that Gene gets out," quips Schmitt at the start of the first EVA. Starting at 6:55 PM and lasting 7 hours and 12 minutes, it proves to be a tiring but productive time. The first task after the walkaround inspection (**2**) is to free the lunar rover, which takes 16 minutes to lock together. The basic vehicle (**1**) minus equip-

2

5

6

7

ment is taken for a short checkout ride by Cernan. The equipment later loaded on the LRV will include the ground controlled television assembly, the lunar communications unit, the high and low gain antennas, a tool pallet and the assorted lunar tools and scientific gear. The flag is set up, and the by-now ritual salutes and pictures are taken. (**3, 4, 7 and 8**) As the immediate area is explored both men do their share of stumbling and fumbling. At one point Schmitt remarks that, "Well, I haven't learned to pick up rocks, which is a very embarrassing thing for a geologist." Small samples of rock ranging in size from ½-inch to 1 inch in diameter are gathered (**5**) with the Lunar Rake. While Schmitt is engaged with the deployment of the ALSEP packages, Cernan attempts a temporary repair on the rear fender of the lunar rover. He had caught it with his hammer and had torn off about ⅓ of it. The repair consisted of taping the broken section back on with adhesive tape. The rover is again taken for a test drive (**6**). A short exploration trip in the rover is undertaken, and Schmitt (**9**) stops to take samples from a large boulder. On the way back the taped splice on the rear fender gives way and part of the fender is lost. The resulting dust thrown up by the wheel causes some worry to both astronauts, as it is feared that the dust will clog their equipment. Capcom promises to work on the problem during the first rest period, for the rover will definitely be needed for the two upcoming EVA tours. The first lunar automotive repair awaits.

3

4

8

9

APOLLO 17 Exploration

When almost five hours into the second EVA, Schmitt and Cernan stop the rover near the rim of the crater Shorty. This point, designated Station 4, is where the astronauts first sight the orange soil deposits. In this panoramic view (**1**) by Cernan, the colored soil is visible on both sides of the rover. The soil coloration and its dispersing pattern around the crater are certain indications of volcanic activity. In center background is Family Mountain, while at left a portion of South Massif is visible. A closeup view (**2**) of a patch of orange soil discovered at Station 4. Bootprints and sample scoops have disturbed the area, while in the background, the crew has placed a gnomon which gives the local vertical reference. A card on the front leg of the tripod gives a color reference comparison. After setting up the gnomon once more, Schmitt (**3**) works with the lunar scoop near a boulder. The rock-strewn soil is almost devoid of color variation in this area. Note Schmitt's Hasselblad camera on its carrier mount. When the

Lunar Roving Vehicle was initially deployed, Gene Cernan took it out for a test drive and subsequently parked it next to *Challenger*. While deploying the ALSEP equipment, Cernan accidentally snagged the right rear fender with his hammer tong, breaking it off. An initial repair with tape did not hold up, and traversing during EVA-1 proved that full fender coverage was necessary due to the quantities of lunar dust thrown up. Astronaut John Young at Mission Control thought up the repair (**4**) seen here while Schmitt and Cernan rested before the present EVA. Here is the first automotive repair job on the moon. Four unused plasticised map sheets are attached to the fender frame with clamps taken from the light for the optical alignment telescope. This jury rig does not fail. Note the coating of dust on the rover from earlier operating. The two explorers are caught together (**5**) by the lunar rover's TV camera. Though their features are indistinguishable, Gene Cernan stands at the left with his arm extended. Next to him is Scientist/Astronaut Harrison Schmitt. "Oops!" Once again (**6**), the rover-mounte RCA television camera catches Schmitt o the lunar surface during EVA-2, this time i a rather embarrassing moment. Losing hi balance in the one-sixth gravity, he head for a fall into the dust. All the surface-rovin astronauts were able to adapt quite rapidl to the low gravity, due in part to excellen training on earth. The deployment and acti- vation of experiments on the moon were al rehearsed many times by the astronauts

4

5

6

APOLLO 17
Exploration

During EVA-3, man's final surface studies of the moon, Schmitt catches Gene Cernan in the corner of this photo (**1**) showing earth with the rover's high-gain antenna aimed directly towards it. In the rear is South Massif. Schmitt (**2**) taps a boulder for samples with his hammer. Though "soft" appearing in photos, certain lunar features prove to be quite hard, and resist the astronauts' attempts at dislodging samples. Cernan's shadow (**3**) projects sharply southward in this spectacular panorama of Taurus-Littrow area. Visible is Schmitt, *Challenger*, an ALSEP and the rover. Note Cernan's footprints. Cernan and Schmitt (**4**) pay their respects to members of the International Youth Science Tour who are visiting MSC during EVA-3. With North Massif in the background (**5**), Schmitt collects samples with the rake. Seated in the rover at Station 9 (**6**) at Van Serg Crater, Schmitt raises his sun visor and settles back into the seat. Visible in this picture are the low-gain antenna and the television camera, which is rotated, its lens facing

1

2

3

4

5

Schmitt. A small red flag (**7**) indicates the location of a Lunar Seismic Probe geophone. Next to it is a familiar gnomon marker, complete with its color reference card. In the background, looking somewhat disorderly, are various ALSEP equipment, while in the distance is the North Massif. With all the ALSEP gear properly deployed, the third and final EVA activities draw to a close. One task remains: the final commemorative of our manned lunar explorations. Gene Cernan (**8**) carefully unveils the small alloy plaque on which is engraved the two hemispheres of our earth with the moon between them. Below it, the dedication to the peaceful intentions under which the astronauts came to the lunar surface, beginning with Apollo 11, and completed here with Apollo 17. At the bottom of the plaque are the signatures of Astronauts Cernan, Evans and Schmitt. Beneath their signatures is the signature of President Richard Nixon. The plaque can be seen in position between rungs of the descent ladder. It is mounted on one of the LM's landing legs. Finally, a last look at the lunar surface (**9**) as seen through the window of *Challenger.* The American flag, hung in Mission Operations Control Room since Apollo 11, has been left on the surface by the crew of Apollo 17. Partly in shadow in the background is North Massif. The flag is partially framed by two of the LM's directional thrusters. On EVA three, both Gernan and Schmitt collect a total of 145 pounds of lunar rocks. They also drive the lunar rover about 11.6 kilometers (some 6 miles)

6

7

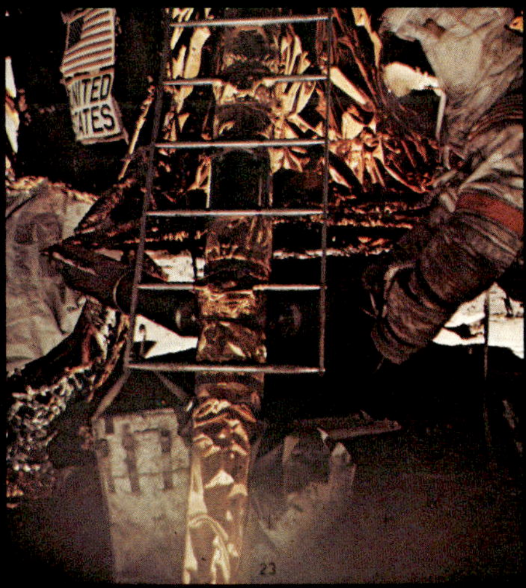

8

9

APOLLO 17
In-flight

During the surface exploration phase, Evans is busy with numerous experiments in lunar orbit. In addition to the cameras in the SIM bay, Evans took many photos using the Hasselblad 500EL/70, such as the one here (**1**) showing Eratosthenes Crater and the southern tip of the Apennine range. His 72 hours of solitude in lunar orbit would soon be broken with the return of Cernan and Schmitt. On the surface, both astronauts are now inside the LM and are making preparations for the ascent phase. The module is pressurized, and both astronauts proceed to doff the extravehicular portions of their space suits. Cernan (**2**) has just removed his helmet. Note the lunar dust that has accumulated on his suit. The danger of dust clogging the equipment was a constant source of worry to both the astronauts and Mission Control. Schmitt (**3**) has stripped down to his constant-wear garment and proudly displays his three-day growth of beard. In the mission operations control room back in Houston various specialists are

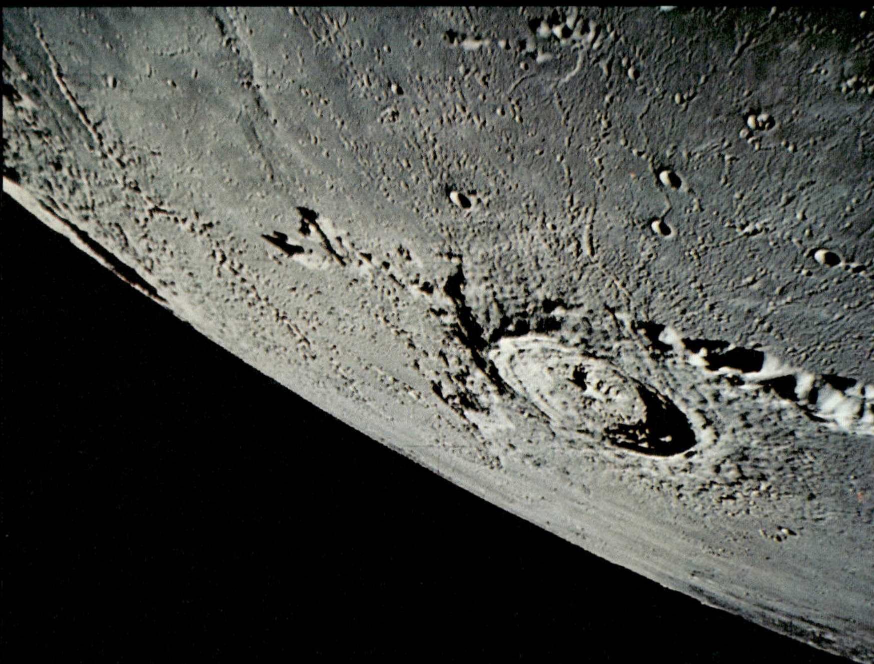

1

2

4

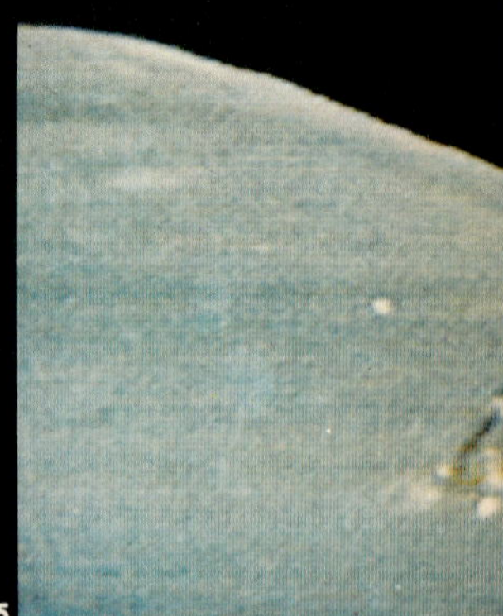

5

monitoring consoles on which are displayed television presentations of data, curves and pictures. These control and monitoring positions were not active during most of the flight, but now they present a picture of intense activity as the pre-liftoff go/no go check list is read off. The room is quiet, and the controller's voice comes on again. "Make your transmissions nice and slow, guys. Let's keep it very cool." Cernan's voice comes on and counts down the remaining seconds. The watching audience in the viewing area sees, on television screens, the spidery-looking Lunar Module standing desolate and lonely against the backdrop of the lunar landscape. Suddenly there is a burst of action (**4**); particles blast away from the invisible flame of the ascent engine, and the upper stage of *Challenger* detaches smoothly and lifts into the lunar sky. (**5**) These pictures are coming from the LRV-mounted TV camera that is remotely controlled from Mission Control in Houston. The data points and communications come rapidly. Cernan's voice calls off the orbital parameters from the on-board computer and the control voice affirms. They are in lunar orbit. The applause starts, in a collective gesture of relief, pride and joy. The ascent stage must achieve orbit within closely defined dimensions to rendezvous with Command/Service Module. *Challenger* sees *America* first, at a distance of 112 nautical miles. Evans, waiting alone in the orbiting CSM *America*, (**6**) says happily, "I don't care what you look like, come on back . . ."

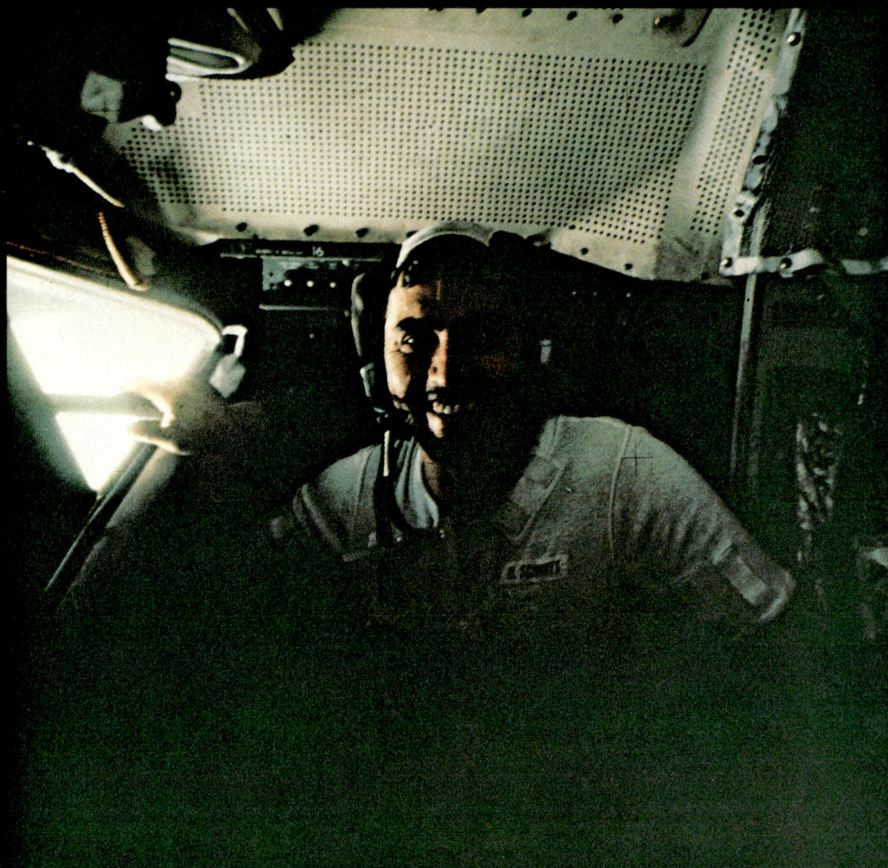

3

6

APOLLO 17 In-flight

The ascent stage of the LM as seen by Ron Evans (**1**) from the Command and Service Module. Gene Cernan's head is just visible in the triangular rendezvous and docking window on the right. Visible on the LM, now within a few meters of the CSM, is the rendezvous radar antenna projecting above the LM's hull. At an angle above the docking window is the VHF In-Flight Antenna. RCS thruster groups are visible at either side of the hull. At the center of the LM, the S-Band In-Flight Antenna casts a shadow across the Tracking Light, which is surrounded by a white bezel. Final docking is achieved after several delicate maneuvers. In final moments prior to docking, the CM and LM ease together (**2**). The docking probe is visible on the nose of the CM. Two docking windows are staggered on either side of the crew hatch of the CM. Seen through the docking window, the CM moves into the shadow of the Lunar Module. While in lunar orbit, Ron Evans captures a magnificent view of earth in crescent phase, rising

2

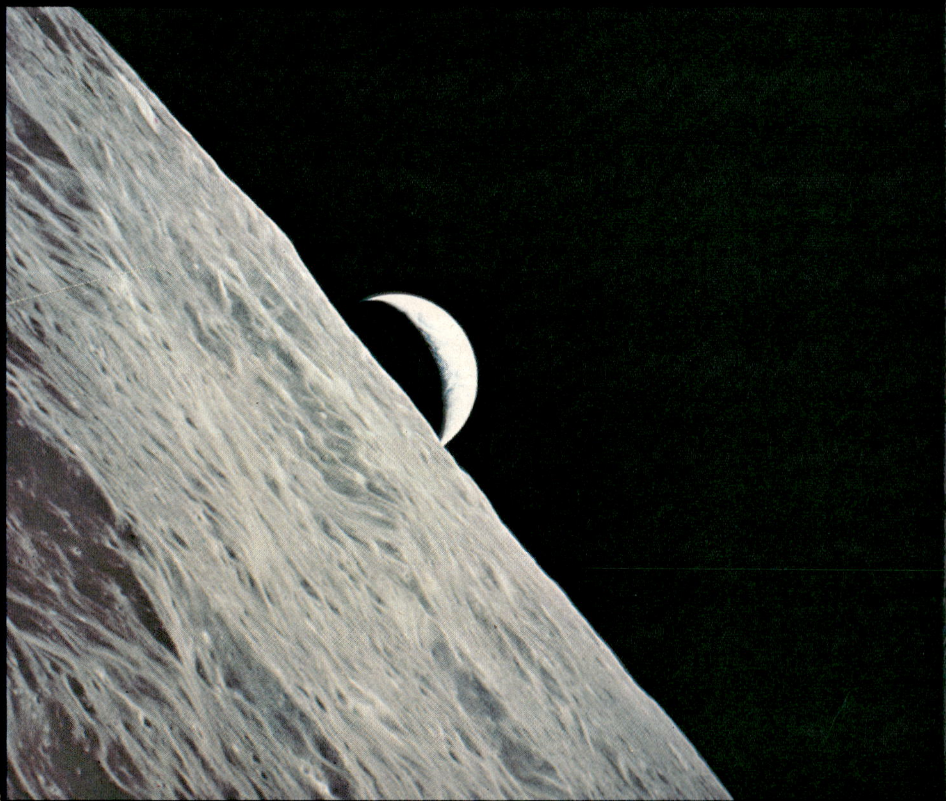

3

over the lunar horizon. (**3**) One of the astronauts commented to Houston regarding awesome scene, a mirror image of what a viewer would see from earth. The battered surface of the moon is razor-clear in all its grey deadness. *Challenger* is now gone forever; separated from *America* for the last time and nudged into a trajectory while orbiting around the back side of the moon, sending it hurtling to the lunar surface. Its impact will be sensed by seismometers on the lunar surface, and the data relayed to earth. Now the final phase of Apollo 17's journey is about to begin. During the 75th orbital revolution of *America*, the trans-earth injection burn is accomplished while on the dark side of the moon. For the second and last time, a TV camera relays pictures of the dark side to earth, the darkside burn pushing 17 into a high orbital escape altitude. This allowed television transmission. As Apollo 17 rounded the moon, and reestablished contact, Cernan radios, "*America* has found some fair winds and following seas, and we're on our way home." But work remains to be carried out, and though most is of an unspectacular nature, Ron Evans' spacewalk (**4**) is not. His EVA is brief; long enough to retrieve the tapes and film cassettes from the SIM bay, which has been deployed during lunar orbit. Freed from the confines of the CM, Evans sings, speaks to his family, and behaves in a now "classic" manner for astronauts in the weightlessness of deep space. SIM bay film cassettes are removed for last time on an Apollo mission.

APOLLO 17 Splashdown

The Apollo series of lunar missions ends on December 17 at 2:24 PM EST with the splash-down of the Apollo 17 Command Module in the Pacific (**1**) about 350 nautical miles southeast of American Samoa. In a way the trip home is anti-climactic. Everything, from the Command and Service Module separation, to the critical re-entry phase goes with clockwork precision. The capsule splashes down within sight of the recovery carrier USS *Ticonderoga.* Within minutes the recovery helicopter is on the scene and the pararescue men in the water. The flotation collar is attached and the rafts moved into position by the hatch. Schmitt emerges first, to be followed by Evans (**3**) and Cernan. After a quick physical examination and a change into flight suits, the three astronauts give an informal briefing on the deck of the *Ticonderoga* (**4**). With the astronauts (left to right) are Captain Norman Green, Commanding Officer of the *Ticonderoga,* and Rear Admiral John Butte Jr., Commander, Task Force 130, Pacific Recovery Area. The

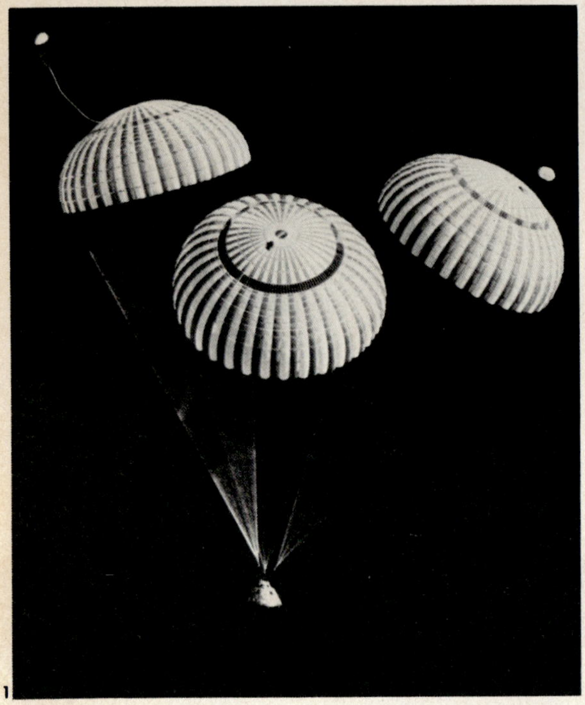

1

2

4